Steffen Rendle

Context-Aware Ranking with Factorization Models

Studies in Computational Intelligence, Volume 330

Editor-in-Chief
Prof. Janusz Kacprzyk
Systems Research Institute
Polish Academy of Sciences
ul. Newelska 6
01-447 Warsaw
Poland
E-mail: kacprzyk@ibspan.waw.pl

Further volumes of this series can be found on our homepage: springer.com

Vol. 308. Roger Nkambou, Jacqueline Bourdeau, and Riichiro Mizoguchi (Eds.)
Advances in Intelligent Tutoring Systems, 2010
ISBN 978-3-642-14362-5

Vol. 309. Isabelle Bichindaritz, Lakhmi C. Jain, Sachin Vaidya, and Ashlesha Jain (Eds.)
Computational Intelligence in Healthcare 4, 2010
ISBN 978-3-642-14463-9

Vol. 310. Dipti Srinivasan and Lakhmi C. Jain (Eds.)
Innovations in Multi-Agent Systems and Applications – 1, 2010
ISBN 978-3-642-14434-9

Vol. 311. Juan D. Velásquez and Lakhmi C. Jain (Eds.)
Advanced Techniques in Web Intelligence, 2010
ISBN 978-3-642-14460-8

Vol. 312. Patricia Melin, Janusz Kacprzyk, and Witold Pedrycz (Eds.)
Soft Computing for Recognition based on Biometrics, 2010
ISBN 978-3-642-15110-1

Vol. 313. Imre J. Rudas, János Fodor, and Janusz Kacprzyk (Eds.)
Computational Intelligence in Engineering, 2010
ISBN 978-3-642-15219-1

Vol. 314. Lorenzo Magnani, Walter Carnielli, and Claudio Pizzi (Eds.)
Model-Based Reasoning in Science and Technology, 2010
ISBN 978-3-642-15222-1

Vol. 315. Mohammad Essaaidi, Michele Malgeri, and Costin Badica (Eds.)
Intelligent Distributed Computing IV, 2010
ISBN 978-3-642-15210-8

Vol. 316. Philipp Wolfrum
Information Routing, Correspondence Finding, and Object Recognition in the Brain, 2010
ISBN 978-3-642-15253-5

Vol. 317. Roger Lee (Ed.)
Computer and Information Science 2010
ISBN 978-3-642-15404-1

Vol. 318. Oscar Castillo, Janusz Kacprzyk, and Witold Pedrycz (Eds.)
Soft Computing for Intelligent Control and Mobile Robotics, 2010
ISBN 978-3-642-15533-8

Vol. 319. Takayuki Ito, Minjie Zhang, Valentin Robu, Shaheen Fatima, Tokuro Matsuo, and Hirofumi Yamaki (Eds.)
Innovations in Agent-Based Complex Automated Negotiations, 2010
ISBN 978-3-642-15611-3

Vol. 320. xxx

Vol. 321. Dimitri Plemenos and Georgios Miaoulis (Eds.)
Intelligent Computer Graphics 2010
ISBN 978-3-642-15689-2

Vol. 322. Bruno Baruque and Emilio Corchado (Eds.)
Fusion Methods for Unsupervised Learning Ensembles, 2010
ISBN 978-3-642-16204-6

Vol. 323. Yingxu Wang, Du Zhang, and Witold Kinsner (Eds.)
Advances in Cognitive Informatics, 2010
ISBN 978-3-642-16082-0

Vol. 324. Alessandro Soro, Vargiu Eloisa, Giuliano Armano, and Gavino Paddeu (Eds.)
Information Retrieval and Mining in Distributed Environments, 2010
ISBN 978-3-642-16088-2

Vol. 325. Quan Bai and Naoki Fukuta (Eds.)
Advances in Practical Multi-Agent Systems, 2010
ISBN 978-3-642-16097-4

Vol. 326. Sheryl Brahnam and Lakhmi C. Jain (Eds.)
Advanced Computational Intelligence Paradigms in Healthcare 5, 2010
ISBN 978-3-642-16094-3

Vol. 327. Slawomir Wiak and Ewa Napieralska-Juszczak (Eds.)
Computational Methods for the Innovative Design of Electrical Devices, 2010
ISBN 978-3-642-16224-4

Vol. 328. Raoul Huys and Viktor K. Jirsa (Eds.)
Nonlinear Dynamics in Human Behavior, 2010
ISBN 978-3-642-16261-9

Vol. 329. Santi Caballé, Fatos Xhafa, and Ajith Abraham (Eds.)
Intelligent Networking, Collaborative Systems and Applications, 2010
ISBN 978-3-642-16792-8

Vol. 330. Steffen Rendle
Context-Aware Ranking with Factorization Models, 2010
ISBN 978-3-642-16897-0

Steffen Rendle

Context-Aware Ranking with Factorization Models

Steffen Rendle
Universität Hildesheim
Wirtschaftsinformatik und
Maschinelles Lernen
Marienburger Platz 22
31141 Hildesheim
Germany
E-mail: srendle@ismll.de

ISBN 978-3-642-42397-0 ISBN 978-3-642-16898-7 (eBook)

DOI 10.1007/978-3-642-16898-7

Studies in Computational Intelligence ISSN 1860-949X

Softcover re-print of the Hardcover 1st edition 2011

Typeset & Cover Design: Scientific Publishing Services Pvt. Ltd., Chennai, India.

Printed on acid-free paper

9 8 7 6 5 4 3 2 1

springer.com

Preface

Context-aware ranking is an important task with many applications like recommender systems or search engines. E.g. in recommender systems items (products, movies, ...) and for search engines webpages should be ranked. In all these applications, the ranking is not global (i.e. always the same) but depends on the context. Simple examples for context are the user for recommender systems and the query for search engines. More complicated context includes time, last actions, etc. In all of these settings, the variables (e.g. customer, product) are defined over large categorical domains and the observations are combinations of the variables (e.g. a user buys a product). The major problem is that typically the observations are very sparse and only positive events are observed.

In this work, we develop a theory of context-aware ranking and methods for solving this task. Our theory brings together tasks like item recommendation and tag recommendation that have been studied mostly isolated before. We first investigate the general problem setting and show how pairwise ranking constraints can be inferred from observations. Then, we develop Bayesian Context-aware Ranking (BCR) which consists of a generic optimization and learning method. BCR is both generic in terms of models and the type of context. The optimization criterion BCR-OPT is the MAP estimator of a parametrized model using the inferred ranking constraints. BCR-LEARN is a learning algorithm that maximizes BCR-OPT and is based on stochastic gradient descent with bootstrap sampling of cases. As the variables are defined over a categorical domain, we propose to use factorization models for parametrizing the dependencies between variable instances. The Tucker decomposition (TD) is a general linear factorization model subsuming parallel factor analysis (PARAFAC). We discuss the problems of TD in terms of runtime complexity for multi-mode problems and show that PARAFAC solves this issue. We show that for our tasks both TD and PARAFAC empirically have bad performance either in runtime or quality, and thus we develop the PITF approach (a special case of PARAFAC) that models pairwise interactions explicitly.

In a second part, we apply our theory of context-aware ranking with factorization models to the tasks of item recommendation, tag recommendation and sequential-set recommendation. For item recommendation and tag recommendation we apply the PITF model with BCR optimization directly. For sequential-set recommendation, we extend the Markov chain model by personalization, i.e. each user can have an own transition graph. As this setting is highly sparse, we model the transition cube with the PITF factorization and show that our factorized, personalized Markov chain subsumes both the common matrix factorization model and a factorized standard (non-personalized) Markov chain. In each of the tasks, we conduct a detailed empirically study and show that our approaches outperform the state-of-the-art methods. Moreover, for the task of graph-based tag recommendation our approach won the 2009 ECML/PKDD Discovery Challenge.

Furthermore, two extensions are studied. First, a non-parametric method for generating time-variant factor models is developed. Here, each factor is decomposed into free parameters and a set of basis functions that are generated from the data using a kernel assumption. Finally, the binary maximum-margin matrix factorization classifier is extended to settings where only one class is observed.

Acknowledgements

I would like to thank my advisor Lars Schmidt-Thieme who has been guiding my research since my master studies in Freiburg. I am also very grateful to my second advisor Takashi Washio for his valuable feedback during my stay in Japan. Without their continuing support and inspiration, this book would not have been possible. Many thanks go to my colleagues and friends: Leandro Balby Marinho, Christoph Freudenthaler, Zeno Gantner, Christine Preisach and Karen Tso. This work is influenced by collaborations and many discussions with them. Finally, I would like to express my profound gratitude to my entire family for their encouragement and unconditional support.

Osaka, July 2010 *Steffen Rendle*

PhD-Thesis: Faculty of Mathematics, Natural Science, Economics and Computer Science; University of Hildesheim

Thesis Committee: Lars Schmidt-Thieme (Chair & Reviewer), Takashi Washio (Reviewer), Klaus Schmid, Tomáš Horváth

Contents

Part IV: Extensions

Part V: Conclusion

Part I
Overview

Chapter 1
Introduction

With the emerging growth of the Internet, a huge amount of information is available to anyone. Even though everything *could* be accessed, the problem is to find relevant information. There are many examples where assistance is needed:

- **Online-Shopping:** Finding the right product within a huge catalogue is time-consuming for a user. Static online-shops organize their products within categories and hierarchies to facilitate browsing. Instead, personalized shops adapt their website to individual customers by factoring in their past actions. This helps the customer to find relevant products faster which leads to an increasing customer satisfaction, a higher purchasing rate and thus more profit. A successful example for such personalized recommender systems is Amazon[1].
- **Tagging:** Tagging is a popular technology in the Web 2.0. Tags allow the user to annotate items/ resources like songs, pictures, bookmarks, etc. with individual keywords. Tagging helps the user to organize his items and facilitate e.g. browsing and searching. But also in the process of tagging (that means annotating the 'right' keywords), assistance is important. Tag recommenders support the tagging process of a user by suggesting him tags that he is likely to use for an item.
- **Search Engines:** Web search is one of the most important tools of the Internet. It helps to find relevant information that is stored in the web. Typically, textual queries are used to search for web pages. The search engine returns a ranked list of pages that matches to the query. Most engines take the location of the user into account. Some engines adapt the results also to the individual user (Sun et al, 2005; Jeh and Widom, 2003).
- **Annotation:** Collaborative creation of content is another recent trend. The online encyclopedia Wikipedia[2] is the most famous example. All content is generated by the visitors and not by a small team of experts. Categorization of articles and links between articles are essential for browsing such large websites. But as the content is created by a large and diverse group of users, when editing content it is hard to find the right categories and links. Tools can help this process by suggesting categories or links.

[1] `http://www.amazon.com/`
[2] `http://en.wikipedia.org/`

S. Rendle: Context-Aware Ranking with Factorization Models, SCI 330, pp. 3–8.
springerlink.com

Table 1.1 Examples for scenarios of context-aware ranking

Scenario	Entity to rank	Context
Online Shopping	products	customer
Tagging	tags	user, bookmark/ song
Search Engine	web pages	search query, user
Wikipedia Annotation	categories	article

In all these examples, the problem is to rank *entities* given a *context*. Table 1.1 gives examples for the entity to rank and the context for the scenarios mentioned before.

In this work, we develop statistical methods that generate such context-aware rankings given observed data. Examples for this observed data are purchases in online shops, clicks on search result lists or already assigned categories for the wikipedia example. As we will discuss in detail in chapter 3, the problem setting is challenging and differs from standard machine learning settings:

1. Instead of classification or regression, we are interested in the less studied task of ranking. Moreover, unlike to the standard ranking literature, not a global ranking has to be found but the rankings should be *context-aware*. That means for each context another ranking is desired.
2. The observations are highly sparse and difficult to interpret. Sparseness means that for most data no observations have been monitored. For example a user has purchased only few products compared to the size of the whole catalogue. The interpretation of the non-bought products is difficult: Not having bought an item does not have to mean that the user dislikes this item but can also mean that he does not know it yet. This is especially crucial as we are mostly interested in ranking among the products a user has not bought yet.
3. The variables (e.g. user, product, web page, tag, ...) are defined over categorical domains with many levels. In contrast to real valued variables, we have no a priori knowledge about the space of these domains. E.g. one does not know a priori if two users are similar or not. This becomes even more crucial with the high sparsity mentioned above.

To solve these issues, we present a data interpretation that generates pairwise preferences for training. We develop a generic optimization criterion from the maximum a posteriori estimator of the pairwise interpretation. For learning model parameters, we present a gradient descent based optimization algorithm. We tackle the sparsity with factorization models that find latent representations of variable instances. These models allow to propagate information over variables, such that two 'similar' instances influence each other. Both the optimization framework and the proposed factorization models can solve problems of any number of modes. That means unlike traditional recommender systems that only work on two modes (e.g. user and item), our approach works also on problems with more modes (e.g. user, items, location, mood).

We provide case studies for item recommendation, tag recommendation and sequential set recommendation (item recommendation with time). Here, we compare our approach to state-of-the-art methods in each of these fields. It is important to note that our method of context-aware ranking is not limited to these applications. But within this book, we can only cover a fraction of all possible scenarios. We are confident that our method also works well in other applications and we are planing to continue this study in future work.

1.1 Overview

This book is organized in the three main parts of theory and application of context-aware ranking as well as general extensions.

1.1.1 Theory of Context-Aware Ranking

The first part deals with the general theory of context-aware ranking. In three chapters we discuss the problem setting, the optimization approach and the modelling with factorization models:

- **Chapter 3: Ranking from Incomplete Data**
  In this chapter, we develop a general method for context-aware ranking. We start with a detailed analysis of the observed incomplete data of a sparse relation over categorical variables. Then we formalize the task of context-aware ranking and show how pairwise preferences can be generated for obtaining training data. Afterwards we describe how a ranking relation can be expressed by a real valued function or by a tensor in the case of finite, categorical variables. Finally, we discuss evaluation metrics for ranking tasks.
- **Chapter 4: Learning Context-Aware Ranking**
  This chapter introduces the framework for Bayesian context-aware ranking (BCR). First, we derive the optimization criterion BCR-OPT which is the maximum a posteriori estimator. For optimizing models towards this criterion, we introduce the learning algorithm BCR-LEARN that is based on stochastic gradient descent with bootstrap sampling of training cases. We conclude with a comparison of BCR-OPT to other optimization criteria like element-wise losses and the area under the ROC curve.
- **Chapter 5: Factorization Models**
  As model, we propose factorization approaches which can model the latent interactions between variables. We discuss the Tucker decomposition (TD) and the PARAFAC model. From this we derive our pairwise interaction model (PITF) which is a special case of both TD and PARAFAC. We discuss the expressiveness and complexity of these models in detail.

1.1.2 Application of Context-Aware Ranking

In the second part, we apply our theory of context-aware ranking to three scenarios and compare our approaches to state-of-the-art methods within these fields.

- **Chapter 6: Item Recommendation**
  Item recommendation is the most well studied recommendation task for ranking. It is a two mode problem over users and items. Online shopping is an example for item recommendation. K-nearest neighbor and matrix factorization are the two most popular approaches for this task. We show how to apply our BCR optimization to both of these models. We compare BCR optimized models empirically to two state-of-the-art approaches: i.e. cosine-similarity and weighted least-square optimization.
- **Chapter 7: Tag Recommendation**
  Tag recommendation is a rather new field of study. Nevertheless it has already attracted a lot of research and many methods to solve this specific tasks have been proposed. In this chapter, we introduce the application of tag recommendation in detail and show how our context-aware ranking theory can be applied there. As indicated before, tag recommendation is a three mode problem over users, items and tags. We show empirically that our approach outperforms the current state-of-the-art methods including Folkrank and HOSVD both in runtime and quality. Furthermore our method was compared to many other approaches in the ECML/ PKDD Discovery Challenge 2009 where we achieved the best quality.
- **Chapter 8: Sequential Set Recommendation**
  Time is a variable that is easy to track in almost all scenarios. In this chapter, we investigate the three mode problem of item recommendation with time. In comparison to the categorical variables that we have used so far, time is real valued. Thus, we have to treat it differently. Here, we examine sequences of shopping carts which reflects the sequential nature of time. For modelling, we use Markov chains. But instead of using standard chains, we introduce personalized (context-aware) chains – i.e. per user one chain. To overcome the sparsity problem, we factorize the chains. Empirically, we show that our new method of factorized personalized Markov chains outperforms both standard Markov chains and time-invariant factorization models.

1.1.3 Extensions

The third part covers two extensions that fall not directly under the theory of context-aware ranking.

- **Chapter 9: Time-variant Factorization Models**
  Instead of modelling the qualitative/ sequential aspect of time like in chapter 8, we model the time quantitatively inside the factors. Therefore, we make the Tucker decomposition time-variant by modelling each factor with a time-variant function. This function itself is factorized into basis functions and free parameters that should be estimated. Instead of choosing the basis functions fixed, we sample them from the observed data using a kernel approach. Finally, we evaluate time-variant models with Gaussian and exponential kernels on synthetic and real-world data sets. Note that this whole chapter is a general study of time-variant factors and not limited to context-aware ranking.

- **Chapter 10: One-Class Matrix Factorization**
  In the last chapter, we investigate a binary classification task over two modes where only one class is observed. Maximum margin matrix factorization (MMMF) is known to be a successful classifier for binary classification tasks over two modes but it is unclear how to apply it to one-class problems. Support vector machines (SVM) are another maximum-margin classifier that have already been applied for one-class problems. We transfer these ideas from one-class SVM to MMMF and propose one-class/ 1C-MMMF. Furthermore we extend it for cases where information about the prior class distribution is available to 1C-prior MMMF.

1.2 Contributions

The core contributions of this book are:

1. We develop a unified **theory of context-aware ranking** that subsumes several recommendation tasks including item, tag and context-aware recommendation.
2. **BCR optimization and learning** is proposed as a generic optimization framework.
3. **Factorization models** are used for modelling and we develop the **PITF** model for sparse problems.
4. **Factorizing Personalized Markov Chains (FPMC)** is introduced as an extension of Markov chains that also allows parameter estimation under sparsity.
5. We conduct **empirical studies** on the task of item recommendation, tag recommendation and sequential set recommendation.
6. **Time-aware factor models** are developed as a time-variant extension of general factorization models.
7. **One-class matrix factorization** with prior regularization is proposed to solve large scale problems with balanced classes.

1.3 Published Work

This book generalizes and builds on the following publications:

- Rendle and Schmidt-Thieme (2008), *Online-updating regularized kernel matrix factorization models for large-scale recommender systems*, in RecSys 08: Proceedings of the 2008 ACM conference on Recommender systems, ACM.
- Rendle, Marinho, Nanopoulos, and Schmidt-Thieme (2009), *Learning optimal ranking with tensor factorization for tag recommendation*, in KDD 09: Proceeding of the 15th ACM SIGKDD international conference on Knowledge discovery and data mining, ACM, New York, NY, USA.
- Rendle, Freudenthaler, Gantner, and Schmidt-Thieme (2009), *BPR: Bayesian personalized ranking from implicit feedback*, in Proceedings of the 25th Conference on Uncertainty in Artificial Intelligence (UAI 2009).

- Rendle and Schmidt-Thieme (2009), *Factor models for tag recommendation in bibsonomy*, in Proceedings of the ECML-PKDD Discovery Challenge Workshop. **ECML/PKDD 2009 Best Discovery Challenge Award**
- Gantner, Freudenthaler, Rendle, and Schmidt-Thieme (2009), *Optimal ranking for video recommendation*, in Personalization in Media Delivery Platforms Workshop at the International ICST Conference on User Centric Media (PerMeD 2009).
- Rendle and Schmidt-Thieme (2010), *Pairwise interaction tensor factorization for personalized tag recommendation*, in Proceedings of the Third ACM International Conference on Web Search and Data Mining (WSDM 2010), ACM. **WSDM 2010 Best Student Paper Award**
- Rendle, Freudenthaler, and Schmidt-Thieme (2010), *Factorizing personalized markov chains for next-basket recommendation*, in WWW 10: Proceedings of the 19th international conference on World wide web, ACM, New York, NY, USA. **WWW 2010 Best Paper Award**

References

Gantner, Z., Freudenthaler, C., Rendle, S., Schmidt-Thieme, L.: Optimal ranking for video recommendation. In: Personalization in Media Delivery Platforms Workshop at the International ICST Conference on User Centric Media (PerMeD 2009) (2009)

Jeh, G., Widom, J.: Scaling personalized web search. In: WWW 2003: Proceedings of the 12th International Conference on World Wide Web, pp. 271–279. ACM, New York (2003)

Rendle, S., Schmidt-Thieme, L.: Online-updating regularized kernel matrix factorization models for large-scale recommender systems. In: RecSys 2008: Proceedings of the 2008 ACM Conference on Recommender Systems, pp. 251–258. ACM, New York (2008)

Rendle, S., Schmidt-Thieme, L.: Factor models for tag recommendation in bibsonomy. In: Proceedings of the ECML-PKDD Discovery Challenge Workshop (2009)

Rendle, S., Schmidt-Thieme, L.: Pairwise interaction tensor factorization for personalized tag recommendation. In: WSDM 2010: Proceedings of the third ACM International Conference on Web Search and Data Mining, pp. 81–90. ACM, New York (2010)

Rendle, S., Freudenthaler, C., Gantner, Z., Schmidt-Thieme, L.: BPR: Bayesian personalized ranking from implicit feedback. In: Proceedings of the 25th Conference on Uncertainty in Artificial Intelligence (UAI 2009) (2009)

Rendle, S., Marinho, L.B., Nanopoulos, A., Schmidt-Thieme, L.: Learning optimal ranking with tensor factorization for tag recommendation. In: KDD 2009: Proceeding of the 15th ACM SIGKDD International Conference on Knowledge Discovery and Data Mining. ACM, New York (2009)

Rendle, S., Freudenthaler, C., Schmidt-Thieme, L.: Factorizing personalized markov chains for next-basket recommendation. In: WWW 2010: Proceedings of the 19th International Conference on World Wide Web, pp. 811–820. ACM, New York (2010)

Sun, J.T., Zeng, H.J., Liu, H., Lu, Y., Chen, Z.: Cubesvd: a novel approach to personalized web search. In: WWW 2005: Proceedings of the 14th International Conference on World Wide Web, pp. 382–390. ACM, New York (2005)

Chapter 2
Related Work

In this chapter, we introduce the general related work for context-aware ranking with factorization models. Related work on specific issues like tag recommenders, Markov chains, etc. is discussed in detail in the corresponding chapters. Here, we discuss three general topics. The first one is recommender systems because the standard task of personalized item recommendation (a two mode problem) can be seen as context-aware ranking where the context is the user. Nevertheless in recommender systems, the term 'context' is usually used only for cases with at least three modes and furthermore the first mode is typically assumed to be the user. Thus, in the discussion about recommender systems we stick to the definition within the recommender community and use the term *context-aware recommender system* only for ranking problems with at least three modes. In contrast to this, in this book we use the term *context-aware ranking* for any number of modes. Secondly, we investigate factorization models on which our proposed approach is based. Finally, we discuss the literature about ranking in general and context-aware ranking in particular.

2.1 Recommender Systems

Traditionally, recommender systems are designed for two-mode problems. The task is to predict how much a user likes an item. In context-aware recommender systems, additional modes are available such as time, mood, etc.

Recommender settings can be distinguished into rating prediction (regression) and item recommendation (ranking). Both settings differ in the type of observations that are available. For rating prediction the observations are explicitly given real values. Item recommendation problems usually have only implicit binary feedback – where only positive feedback is observed. The later problem is the more challenging one as the interpretation of the observation is difficult. Furthermore, it appears more often in practice because implicit feedback is easy to gather – e.g. almost every web server records implicit behavior in log files by default. This book focus on the recommendation problem from implicit feedback.

S. Rendle: Context-Aware Ranking with Factorization Models, SCI 330, pp. 9–15.
springerlink.com

2.1.1 Two-Mode Recommender Systems

Recommender systems are a well-studied field with many different approaches. One of the most popular approaches is k-nearest-neighbor collaborative filtering (Sarwar et al, 2001). Recently, matrix factorization methods have become very popular because of their success on the Netflix challenge (Srebro et al, 2005; Koren, 2008). The item recommendation problem can also been transformed into a multi-class classification problem and standard binary classifiers like SVMs can be applied using the 1-vs-1 or 1-vs-rest scheme (Schmidt-Thieme, 2005). Other well-known approaches for recommender systems are Boltzmann machines (Salakhutdinov et al, 2007) or the PLSA (probabilistic latent semantic analysis) model (Hofmann, 2004). We will discuss these two-mode recommender systems in more detail in chapter 6.

2.1.2 Context-Aware Recommender Systems

In comparison to the vast literature in traditional recommender systems over two modes, context-aware recommender systems over more modes have attracted much less attention. Moreover, only simple methods have been presented so far for context-aware recommender systems. The approaches for context-aware recommenders can be categorized into three types: (1) contextual pre-filtering, (2) contextual post-filtering and (3) contextual modelling.

2.1.2.1 Contextual Pre-filtering

Given a context, in contextual pre-filtering a relevant subset of the observations of the past is generated. Then, a standard two-mode recommender method can be applied on this subset – both for training and prediction. The advantage of this approach is that any traditional two-mode method can be used because the training set is made context-aware and thus the method has not to be adjusted. But this approach has two important limitations:

1. For each context, an own recommender model has to be applied. For lazy methods without a training phase (like typical cosine kNN) this is no problem. But for parametrized models (that are known to provide better quality) like factorization models, Boltzmann machines, etc. for each subset a model has to be learned. This is not feasible for a huge number of context.
2. By creating a subset of the training data, a lot of information is withheld from the recommender system. This is especially important as we deal with a highly sparse setting. Furthermore the skipped data is supposed to contain important information because observations between two context are not completely independent. To tackle the issue of sparsity, Adomavicius et al (2005) suggest to generate a broader subset that includes data not only corresponding exactly to the given context but also to other related context.

2.1.2.2 Contextual Post-filtering

In contrast to pre-filtering, post-filtering takes the contextual information at the end into account. First, all contextual information is discarded and a traditional two-mode recommender system is applied. For predicting, the learned two-mode model is used without contextual information. Context-awareness is ensured by post-processing the recommender list. Panniello et al (2009) suggest to use a weighting or filtering approach. For both approaches a contextual probability for the entity is computed. In the weighting approach, the recommender list is reordered by multiplying with this probability. The filtering approach removes entities with a probability that is smaller than a certain threshold.

An advantage of contextual post-filtering is that in the first step any recommender system can be used. And in comparison to pre-filtering only one model has to be built which makes it more applicable. An open issue for post-filtering is how to obtain the contextual probability. Panniello et al (2009) use a simple estimator by counting occurrences of the item in identical context of users in the neighbourhood. This estimator will be unreliable in sparse settings, because it matches for exact context and items. Again a generalization of the context and/ or item might help.

Panniello et al (2009) have empirically compared the performance of pre- and post-filtering. Their results indicate that neither of the two of them outperform the other.

2.1.2.3 Contextual Modelling

Instead of using two-mode recommender systems and applying just a pre/post-processor, in contextual modelling the recommender system uses context information directly in the model.

Adomavicius et al (2005) present a multidimensional model based on OLAP cubes. As sparsity is a main problem for generating estimations in context-aware settings, aggregation of dimensions is used to generate more reliable estimations. In total, this approach is rather simple because it does not subsume any strong (recommender) model. Another approach is to apply SVMs where the context is part of the feature space (Oku et al, 2006). In general, applying standard classifiers does not scale because the categorical variables have to be encoded as many binary variables resulting in a huge dataset even for mid-sized problems.

All the methods that we develop in this work can be classified as contextual modelling. A strong point of our proposed approach is that it subsumes the best performing methods for two-mode item recommendation and three-mode tag recommendation.

2.2 Factorization Models

Our context-aware models are based on factorization models. As we are dealing with categorical variable domains, the problem can be seen as predicting the entries of a multi-mode tensor. With factorization models, each variable is described by a vector of (latent) variables which are called the factors. The entries of a multi-mode tensor can then be constructed by combining the factors. There are two things to consider when applying factorization models: (1) How the factors interact – this is determined by the model structure/ equation. (2) How the factors are obtained – this is defined by the optimization criterion.

Tucker (1966) suggest to factorize the multi-mode tensor into a smaller core tensor and one factor matrix for each mode. Higher-order singular value decomposition (HOSVD) is one method for obtaining the factors (Lathauwer et al, 2000). HOSVD corresponds to a least-square optimization on a tensor without missing values. Parallel factor analysis (PARAFAC) (Harshman, 1970; Carroll and Chang, 1970) is a special case of the Tucker decomposition (TD) where the core tensor is diagonal. The advantage of PARAFAC over TD is that the model equation has no nested sums and thus is much faster. In the two-mode case PARAFAC corresponds to matrix factorization (MF). Singular value decomposition (SVD) is well-known method for estimating factors of a two-mode PARAFAC/ MF model. Analogously to HOSVD, SVD also optimizes for least-square and does not allow missing values. Another approach is to learn MF with a sparse optimization and ridge regression terms as regularization. This has been introduced as maximum-margin matrix factorization (Srebro et al, 2005) using the hinge loss.

For rating prediction in two-mode recommender systems, sparse matrix factorization with regularization and least-square optimization is known to be one of the best approaches (Koren, 2008). Salakhutdinov and Mnih (2008) have extended this to Bayesian Probabilistic Matrix Factorization where the parameters are learned with Markov Chain Monte Carlo (MCMC). For the three-mode problem of tag recommendation, Symeonidis et al (2008) have used HOSVD. To apply this, the missing values have been imputed with 0.

We will discuss factorization models in detail in chapter 5. The optimization is discussed in chapter 4. Applications and comparisons to state-of-the-art models are provided in part III.

2.3 Ranking

Besides recommender systems, there is several work on general ranking. First, we discuss approaches that learn models for optimal ranking. Secondly, we investigate context-aware approaches for ranking.

2.3.1 Global Ranking

There are several approaches that try to optimize models for ranking. Both Kondor et al (2007) and Huang et al (2008) model distributions over permutations. Burges et al (2005) optimize a neural network model for ranking using gradient descent. All these approaches learn only one ranking – i.e. they are not context-aware. In contrast to this, our models are collaborative models that learn context-aware rankings, i.e. one individual ranking per context. In the application part of this work, we show empirically that in our settings it is important to take context into account and that our BCR optimized factorization models outperform even the upper bound for non context-aware ranking.

One way to make the global ranking models context-aware is to apply the idea of contextual-prefiltering (see section 2.1.2.1). That means for each context an individual model is used. For example, for item recommendation each user has an own model or for tag recommendation each post (user-item combination) has an own model. Each individual model is learned from the subset of the training data that matches to the model's context. Obviously, the training data for each model is very small which results in poor parameter estimates. The reason is that the parameters for each model are independently and thus no inference across context is possible. Moreover this approach is not able to learn for unobserved context, e.g. an unobserved post. In contrast to this, the factorization approach that we propose in this book does not require that the values of a context have been jointly observed – instead it can infer across context.

2.3.2 Context-Aware Ranking

On the other hand there are context-aware ranking approaches. Agrawal et al (2006) investigate context-sensitive ranking. But their problem setting differs substantially from ours. In addition to a dataset like in our case, they assume that a set of contextual preferences is given in advance (alternatively they can also be learned, e.g. by association rule mining). A contextual preference is a binary relation over two variable instances given a context. The type of context they investigate is given by a conjunction of equality constraints. Thus it is possible that the context is sparse, i.e. some of the variables are not defined. The task they solve is for a given query (e.g. SQL) to take the contextual preferences into account and to rank the resulting tuples. If we would apply this to our sparse problem setting, the where-clause of the query would contain the complete context. As we are dealing with very sparse settings, the selection would typically be empty, because it is very unlikely that there are observations for exactly this context. Furthermore, we are not interested to rank for context that has been observed already but rather in ranking for non-observed context. In total, the problem settings of our work and in (Agrawal et al, 2006) are too different and thus their method does not make sense in our setting and vice versa, our approach is supposed to perform bad in their setting.

Haveliwala (2003) describes a context-sensitive extension of the famous Pagerank (Brin and Page, 1998) algorithm. The idea is to generate a context-aware Pagerank in three steps: (1) A small set of 'topics' (e.g. 16 in his experiments) is selected and one Pagerank for each of these topics is generated. (2) A probabilistic classifier is learned to map a context to the topics. For this they use a Naive Bayes classifier. (3) The context-sensitive Pagerank is now the weighted average of the topic Pageranks where the probabilities of the topic classifier are used as weights. In our settings no topics are given in advance. Thus one could model the topics as latent topics. The factorization dimensions of our factorization models can be seen as a kind of latent topics. Instead of finding latent topics just for the entities to rank, factorization models also generate factors for the variables in the context. Furthermore, our factorization models optimize all 'topics'/ factors jointly and also do not need to learn any mapping from topics to rankings.

References

Adomavicius, G., Sankaranarayanan, R., Sen, S., Tuzhilin, A.: Incorporating contextual information in recommender systems using a multidimensional approach. ACM Transactions on Information Systems 23(1), 103–145 (2005)

Agrawal R, Rantzau R, Terzi E (2006) Agrawal, R., Rantzau, R., Terzi, E.: Context-sensitive ranking. In: SIGMOD 2006: Proceedings of the 2006 ACM SIGMOD international conference on Management of data, pp. 383–394. ACM, New York (2006)

Brin, S., Page, L.: The anatomy of a large-scale hypertextual web search engine. In: WWW7: Proceedings of the Seventh International Conference on World Wide Web, vol. 7, pp. 107–117. Elsevier Science Publishers B. V, Amsterdam (1998)

Burges, C., Shaked, T., Renshaw, E., Lazier, A., Deeds, M., Hamilton, N., Hullender, G.: Learning to rank using gradient descent. In: ICML 2005: Proceedings of the 22nd International Conference on Machine Learning, pp. 89–96. ACM Press, New York (2005)

Carroll, J., Chang, J.: Analysis of individual differences in multidimensional scaling via an n-way generalization of eckart-young decomposition. Psychometrika 35, 283–319 (1970)

Harshman, R.A.: Foundations of the parafac procedure: models and conditions for an 'exploratory' multimodal factor analysis. UCLA Working Papers in Phonetics, 1–84 (1970)

Haveliwala, T.H.: Topic-sensitive pagerank: A context-sensitive ranking algorithm for web search. IEEE Transactions on Knowledge and Data Engineering 15(4), 784–796 (2003)

Hofmann, T.: Latent semantic models for collaborative filtering. ACM Trans. Inf. Syst. 22(1), 89–115 (2004)

Huang, J., Guestrin, C., Guibas, L.: Efficient inference for distributions on permutations. In: Platt, J., Koller, D., Singer, Y., Roweis, S. (eds.) Advances in Neural Information Processing Systems, vol. 20, pp. 697–704. MIT Press, Cambridge (2008)

Kondor, R., Howard, A., Jebara, T.: Multi-object tracking with representations of the symmetric group. In: Proceedings of the Eleventh International Conference on Artificial Intelligence and Statistics, San Juan, Puerto Rico (2007)

Koren, Y.: Factorization meets the neighborhood: a multifaceted collaborative filtering model. In: KDD 2008: Proceeding of the 14th ACM SIGKDD International Conference on Knowledge Discovery and Data Mining, pp. 426–434. ACM, New York (2008)

Lathauwer, L.D., Moor, B.D., Vandewalle, J.: A multilinear singular value decomposition. SIAM J. Matrix Anal. Appl. 21(4), 1253–1278 (2000)

Oku, K., Nakajima, S., Miyazaki, J., Uemura, S.: Context-aware svm for context-dependent information recommendation. In: MDM 2006: Proceedings of the 7th International Conference on Mobile Data Management, p. 109. IEEE Computer Society, Washington (2006)

Panniello, U., Tuzhilin, A., Gorgoglione, M., Palmisano, C., Pedone, A.: Experimental comparison of pre- vs. post-filtering approaches in context-aware recommender systems. In: RecSys 2009: Proceedings of the third ACM conference on Recommender systems, pp. 265–268. ACM, New York (2009)

Salakhutdinov, R., Mnih, A.: Bayesian probabilistic matrix factorization using Markov chain Monte Carlo. In: Proceedings of the International Conference on Machine Learning, vol. 25 (2008)

Salakhutdinov, R., Mnih, A., Hinton, G.: Restricted boltzmann machines for collaborative filtering. In: ICML 2007: Proceedings of the 24th International Conference on Machine Learning, pp. 791–798. ACM, New York (2007)
Sarwar, B., Karypis, G., Konstan, J., Reidl, J.: Item-based collaborative filtering recommendation algorithms. In: Proceedings of the 10th International Conference on World Wide Web, pp. 285–295. ACM Press, New York (2001)
Schmidt-Thieme, L.: Compound classification models for recommender systems. In: IEEE International Conference on Data Mining (ICDM 2005), pp. 378–385 (2005)
Srebro, N., Rennie, J.D.M., Jaakola, T.S.: Maximum-margin matrix factorization. In: Advances in Neural Information Processing Systems, vol. 17, pp. 1329–1336. MIT Press, Cambridge (2005)
Symeonidis, P., Nanopoulos, A., Manolopoulos, Y.: Tag recommendations based on tensor dimensionality reduction. In: RecSys 2008: Proceedings of the 2008 ACM Conference on Recommender Systems, pp. 43–50. ACM, New York (2008)
Tucker, L.: Some mathematical notes on three-mode factor analysis. Psychometrika 31, 279–311 (1966)

Part II
Theory

Chapter 3
Ranking from Incomplete Data

The most common classification setting is to predict labels from real-valued vectors, e.g. logistic regression or Support Vector Machines (SVM) are designed for this purpose. Our task differs from this: (1) The variables in our settings are defined over categorical domains with very many levels and there is no a priori knowledge about the space the variable instances lie in. (2) The observed data is highly sparse, non-trivial to interpret and it makes statements rather about pairs of instances than about a single instance. (3) The prediction problem is to rank the instances of one variable given an instance vector (the 'context') of the other variables. As the ranking should depend on the given instance vector, it is not a global ranking but a context dependent one. In this chapter, we discuss these three issues and develop a theory for context-aware ranking.

First, we formalize the problem in an abstract way and discuss the problem of sparsity in the observations. Sparsity means that only for little instances feedback is present. Furthermore, this feedback is always positive, i.e. no negative feedback is directly observed. The interpretation of the non-observed instances is not trivial as they contain both the negative instances and the missing positive instances which should be found and ranked high.

Secondly, we introduce context-aware ranking which is formalized as finding a total order on one variable given an instance of all other variables. Then we show how training data for the ranking can be generated from the observed (positive) instances. This is illustrated for the task of product recommendation and tag recommendation.

Finally, we show how modelling rankings can be reformulated as modelling real valued functions. This is important because modelling all constraints of a total order is complicated whereas real valued functions are trivially transitive and connex. We proof that every ranking can be expressed as a real valued function, but also that there are many functions to represent one ranking. All methods in this work are based on modelling such real valued functions instead of modelling a ranking directly.

We conclude this chapter by introducing metrics to measure the empirical quality of a predicted ranking. These measures will be used later to evaluate our proposed ranking methods.

S. Rendle: Context-Aware Ranking with Factorization Models, SCI 330, pp. 19–37.
springerlink.com

In the next chapter, we will apply the ideas of this chapter and develop an optimization criterion and learning algorithm for y. Afterwards, factorization models are introduced to represent y and to overcome the sparsity problem.

3.1 Sparse Observations

First, we investigate the problem setting. We formalize the problem in terms of variables, relations over variables and instances of variables / relations. Then we compare the problem setting with standard classification and regression problems. The main problem we discuss in detail is how to interpret the observed sparse data.

We use two examples throughout this chapter for illustrating the problems: (1) An online shop where customer buy products. The task is to recommend products to a customer. (2) A social website with tagging capabilities like Last.fm where users annotate items (e.g. songs) with tags. Each user can give individual keywords.

3.1.1 *Variables*

Let $X_1, \ldots, X_m$ be domains over which m variables are defined. In this work, the domains are categorical that means a priori there is no relationship between elements in X like an order nor are there any mathematical operations like '+' defined over two elements in X. The only exception that we discuss is time where the domain is $\mathbb{R}$ with all properties like ordering etc.

Furthermore, we define the space $\mathscr{X}$ over the domains:

$$\mathscr{X} := X_1 \times \ldots \times X_m \tag{3.1}$$

We call the vector $\mathbf{x} = (x_1, \ldots, x_m) \in \mathscr{X}$ an instance/ element of $\mathscr{X}$ and m the *mode*.

Examples

- **Online shop:** this example can be formalized with two categorical variables:

$$m = 2$$
$$X_1 = \{customer_1, customer_2, \ldots\}$$
$$X_2 = \{product_1, product_2, \ldots\}$$

Thus, the instance $(customer_1, product_2)$ of the space $\mathscr{X}$ would mean that $customer_1$ has bought $product_2$.

In an extended scenario, we could also model the time with an additional variable over the real numbers:

$$m = 3$$
$$X_1 = \{customer_1, customer_2, \ldots\}$$
$$X_2 = \mathbb{R}$$
$$X_3 = \{product_1, product_2, \ldots\}$$

Here an instance of $\mathscr{X}$ would also state when the customer has bought the item.

- **Tagging:** for the tagging scenario, three variables over the following domains are necessary:

$$m = 3$$
$$X_1 = \{user_1, user_2, \ldots\}$$
$$X_2 = \{item_1, item_2, \ldots\}$$
$$X_3 = \{tag_1, tag_2, \ldots\}$$

- **Search-Engine:** an additional example is a website of a search engine, where visitors use queries to search and then click on links to web pages:

$$m = 3$$
$$X_1 = \{visitor_1, visitor_2, \ldots\}$$
$$X_2 = \{query_1, query_2, \ldots\}$$
$$X_3 = \{webpage_1, webpage_2, \ldots\}$$

3.1.2 Observations

Next, we define the observations that have been made in the past. Statistical methods use these observations to learn regularities and generate recommendations/ rankings. The observations are defined over $\mathscr{X}$ and each instance $\mathbf{x} \in \mathscr{X}$ can occur multiple times in the past.

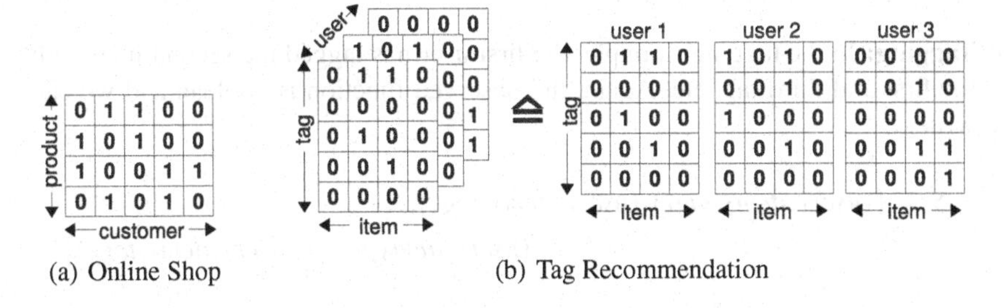

(a) Online Shop (b) Tag Recommendation

Fig. 3.1 The observed data s of the past in an online shop (left) and tagging scenario (right). Each relation *customer* × *product* and *user* × *item* × *tag* can be written as a tensor. For the tagging scenario, the slices of the cube have been written next to each other.

Let s be the function describing the observations in the past:

$$s : \mathscr{X} \to \mathbb{N} \tag{3.2}$$

Set Notation

Mostly, s is a binary function – i.e. $s : \mathscr{X} \to \{0,1\}$. Thus s can also be seen as a set S:

$$S \subseteq \mathscr{X} \tag{3.3}$$

where S is the support of s:

$$S := \sup s := \{x \in \mathscr{X} | s(x) \neq 0\} \tag{3.4}$$

or in our case equivalently:

$$\mathbf{x} \in S :\Leftrightarrow s(\mathbf{x}) > 0 \tag{3.5}$$

If for all instances $\mathbf{x} : s(\mathbf{x}) \leq 1$ then with this transformation no information is lost and we can use sets instead of functions.

Examples

Figure 3.1 shows an example for observed data in the online shopping and tagging scenario.

- **Online shop:** In figure 3.1(a), the first customer has bought the second and third products whereas the second customer has bought the first and last product, etc. As every entry is smaller or equal to 1, we can use the set notation to formalize S which defines the corresponding s:

$$S := \{(customer_1, product_2), (customer_1, product_3), \ldots, (customer_5, product_3)\}$$

- **Tagging:** In the tagging example, the first user has tagged the second item with the first and third tag, etc. Again, the observed function is boolean and we can write:

$$S := \{(user_1, item_2, tag_1), (user_1, item_2, tag_3), \ldots$$
$$\ldots, (user_3, item_4, tag_4), (user_3, item_4, tag_5)\}$$

3.1.3 *Prediction as Classification/Regression*

At first glance, the task of recommending looks like a classification/ regression task where the function s should be estimated. After having learned $\hat{s} : \mathscr{X} \to \mathbb{R}$ one might rank the instances $\mathbf{x}$ by $s(\mathbf{x})$. Learning s for regression can be done by optimizing a model with respect to the least square loss – and for classification e.g. with respect to the hinge loss. There are two problems with this approach:

1. The optimization criterion does not directly reflect the ranking task. That means the optimization is done against another task than the one that is desired.
2. The characteristics of the data are not taken into account. The observed data s is highly sparse with almost no training examples on instance level (per instance $\mathbf{x}$). That makes especially non-observed entries ($s(\mathbf{x}) = 0$) hard to interpret. Furthermore as the non-observed entries are the ones a recommender system typically is interested in recommending, optimizing a model for predicting all of them as zero / negative will result in bad predictions given enough expressiveness of the model. For example, a very simple 1-NN model would perfectly fit the observations ($\forall \mathbf{x} \in \mathscr{X} : \hat{s}(\mathbf{x}) = s(\mathbf{x})$), but obviously this model is useless for recommending. Besides this, the high sparsity in a classification/ regression setting would mean high imbalance – i.e. the majority class ($s(\mathbf{x}) = 0$) hugely dominates the class with positive observations.

We will discuss the problem of sparsity in more detail next. Later, chapter 4 targets the optimization task.

3.1.4 Sparsity

Usually, for most instances $\mathbf{x}$ there are no observations in the past and thus:

$$\sum_{\mathbf{x} \in \mathscr{X}} \delta(s(\mathbf{x}) > 0) = |\sup s| \ll |\mathscr{X}| \tag{3.6}$$

or in set notation:

$$|S| \ll |\mathscr{X}| \tag{3.7}$$

And often, if an instance $\mathbf{x}$ has been observed, it has rarely been observed multiple times – e.g. a customer usually buys a certain product (e.g. a book) just once and a user chooses a certain tag for an item just once. That means for almost all instances $\mathbf{x}$ there is usually no observation, i.e. $s(\mathbf{x}) = 0$. E.g. in chapter 7, for the (densified) Last.fm example, 99.998% of the values are unobserved; for the ECML/PKDD Challenge datasets that is even more: 99.99993%! That means building regression or classification models for approximating s directly is likely to fail because the simple estimator $\forall \mathbf{x} : \hat{s}(\mathbf{x}) = 0$ would be almost perfect in terms of accuracy or square loss. But this estimator would obviously be useless for ranking.

Secondly, it is crucial to note that if an instance $\mathbf{x}$ has not been observed ($s(\mathbf{x}) = 0$) this does not mean that it won't be observed in the future. Quite the contrary: these instances are exactly those instances from which one has to choose the next actions. This becomes more clear by the example of online shopping: even though the first customer has not bought the first and forth item yet (see fig. 3.1(a)), it does not mean that he will never buy them, but instead the recommender system should decide in which of the two of them he is more interested in[1]. Again, a classifier that fits s perfectly (e.g. the 1-NN mentioned above) could not do any ranking on these unobserved pairs as it predicts them all equally as 0.

The tagging example (see fig. 3.1(b)) makes a related problem obvious: even though the first user has not given any tag to the first item yet, this does not mean that the user has expressed that from his point of view all these tags are bad for describing the item. A better interpretation would be that the user has not yet tagged the item at all – e.g. because he is not interested in the item or has not found it yet. That means it is a (weak) statement rather about the user-item pair than about the tags.

Another reason why sparsity is a problem, are the categorical variables. Imagine a classification example where two dimensional points over the reals should be classified into the positive and negative class. When we draw cases from the positive class, it is also the case that no instance is observed twice (assuming that the distribution is continuous). But this scenario with reals is different from our categorical domains, as the cases are defined over $\mathbb{R}^2$. Assume we have observed that $(1.0, 1.2)$ and $(1.2, 1.0)$ are positive, then it makes sense to follow that $(1.1, 1.1)$ is also positive. This can be done because a priori knowledge over $\mathbb{R}^2$ is present, such as $<$ relations or (e.g. euclidean) distance – this corresponds to a kind of smoothness assumption. In our case with categorical variables, this knowledge is not given and thus it is much harder to make inference over sparse elements in $\mathcal{X}$.

In the following, we will show an alternative way of handling the sparsity. Instead of seeing the problem as an element-wise classification task, we will formulate it as a ranking task based on pairwise classification. We will start with introducing context-aware ranking and then we show, how training data for the ranking can be inferred from s.

3.2 Context-Aware Ranking

Recommendation tasks are rather a ranking than a classification/ regression problem. Instead of predicting one value per single instance $\mathbf{x}$, ranking approaches predict an order over a set of instances. That means, the targets of the instances depend on each other. We start by formalizing a ranking as a strict total order and then extending it to context-awareness. This allows to change the ordering of instances depending on a context. The following section then discusses how to generate training instances for the ranking from s.

3.2.1 *Ranking*

A ranking $\succ$ of the variable instances of a domain X is a strict total order on X. Without loss of generality, we assume that x_m is the target variable to rank.

[1] Empirically, this has been observed in the failing of 1-vs-rest classification models whereas 1-vs-1 models work (Schmidt-Thieme, 2005).

$$\succ \subseteq X_m^2 \tag{3.8}$$

We write:

$$x_a \succ x_b \;:\Leftrightarrow (x_a, x_b) \in \succ \tag{3.9}$$

A strict total order has to be irreflexive, connex and transitive[2]:

$$\forall x_i \in X_m : \neg(x_i \succ x_i) \tag{3.10}$$

$$\forall x_i, x_j \in X_m : x_i \succ x_j \vee x_j \succ x_i \vee x_i = x_j \tag{3.11}$$

$$\forall x_i, x_j, x_k \in X_m : x_i \succ x_j \wedge x_j \succ x_k \Rightarrow x_i \succ x_k \tag{3.12}$$

Because of these axioms, we can define the inverse ranking $\prec$ as:

$$x_a \prec x_b \;:\Leftrightarrow x_b \succ x_a \tag{3.13}$$

3.2.2 *Context*

Instead of finding the same ranking $\succ$ for all context (e.g. all users), the ranking should be context-sensitive – i.e. individual for each context. A context can be seen as the situation under which a decision should be made, e.g. the time, place and an individual person can be the context of ranking products. A *context* $\mathbf{c} \in \mathscr{C}$ is defined as a relation over all the remaining variables

$$\mathscr{C} := X_1 \times \ldots \times X_{m-1} \tag{3.14}$$

And we will write $\mathbf{c}$ for the vector over an instance of $\mathscr{C}$:

$$\mathbf{c} = (x_1, \ldots, x_{m-1}) \in \mathscr{C} \tag{3.15}$$

Furthermore for easier readability, we sometimes write $\mathbf{x}' = (\mathbf{c}, x_m)$ for an instance $\mathbf{x}'$ of $\mathscr{X}$ with $x'_1 = x_1, \ldots, x'_{m-1} = x_{m-1}, x'_m = x_m$. This notation is useful when we want to work with the context $\mathbf{c}$ of an instance $\mathbf{x} \in \mathscr{X}$.

3.2.3 *Context-Aware Ranking*

Now, the task of context-aware ranking is to find for each context $\mathbf{c}$ a ranking $\succ_{\mathbf{c}}$. In total:

$$\succ \subseteq \mathscr{C} \times X_m^2 \tag{3.16}$$

Again, we write:

$$x_a \succ_{\mathbf{c}} x_b \;:\Leftrightarrow (\mathbf{c}, x_a, x_b) \in \succ \tag{3.17}$$

And also each ranking $\succ_{\mathbf{c}}$ has to be a strict total order (see eqs. (3.10)-(3.12)).

[2] Alternatively, one can define a strict total order with transitivity and trichotomy.

Examples

- **Online shop:** In online shopping, the domain that should be ordered are the products. The context is the customer. That means the task is to estimate a personalized ranking of products individually for each customer. E.g. for a customer that usually buys classic novels, books like 'Catch-22' or 'The Old Man and the Sea' will be ranked high whereas for parents usually buying books for their child, books like 'Winnie the Pooh' might rank first.
- **Tagging:** For tagging, the domain to rank are the tags. Here the context is the user and the item. That means for every user-item combination another ranking of tags is generated. Having individual rankings per item (e.g. per song) obviously is important because the ranking of tags for a song by 'The Beatles' should be different from the ranking of tags for classical music e.g. by Beethoven. But furthermore also the user for which the tags are recommended is important because different users tag differently – e.g. their taste and purpose for tags differs even for the same item.
- **Time-awareness:** If time is also monitored, the context should be extended by time. Then the ranking is also time-dependent. E.g. in the online-shopping scenario, one could rank items differently for Christmas than for Easter. Here, the ranking of products would be both customer and time specific.

3.3 Generating Ranking Constraints

In section 3.1.4, we have shown that the observed data s is usually highly sparse and for most instances $\mathbf{x}$ there are no observations ($s(\mathbf{x}) = 0$). Furthermore, we are interested in generating context-aware rankings instead of classification. That means for each context $\mathbf{c}$, an order $\succ_{\mathbf{c}}$ should be found. In the data are no direct observations of $\succ$ but only s is observed. Now, we show how to derive training examples for $\succ$ from the observed data s.

As $\succ$ is a binary relation over $\mathscr{C} \times X_m^2$, the training data on $\succ$ is a function d_s holding information about empirically observed instances of $\succ$:

$$d_s : \mathscr{C} \times X_m^2 \to \mathbb{N} \tag{3.18}$$

Next, we will show how to derive pairwise training data d_s from s.

3.3.1 *Training Data for Rankings*

Training data for rankings is not directly observed. But the observed data s gives for some of the triples $(\mathbf{c}, x_A, x_B) \in \mathscr{C} \times X_m^2$ an indication if $x_A \succ_{\mathbf{c}} x_B$ or $x_B \succ_{\mathbf{c}} x_A$. Imagine the case where a customer $cust_1$ has bought the products p_2 and p_3 but neither product p_1 nor p_4 (figure 3.2, top). This indicates that the customer prefers p_2 over p_1, p_2 over p_4, p_3 over p_1 and p_3 over p_4 (3.2, middle). But no indication whether he likes p_1 or p_4 better nor whether he likes p_2 or p_3 is present. In total,

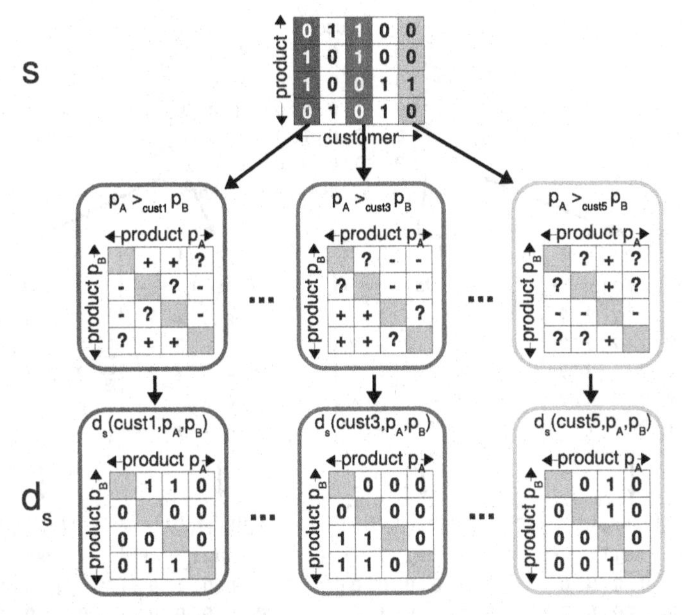

Fig. 3.2 Online Shopping: generating training examples d_s for $\succ$ from s. The topmost matrix shows the observed data in the past (see figure 3.1(a)). Below there is one matrix for each customer that indicates whether or not one can infer that the customer prefers one product over the other. '+' indicates that p_A is preferred over p_B, '-' states that p_B is preferred over p_A and '?' means that no preference can be deduced. On the bottom is the inferred training data d_s for the relation $\succ$.

four training cases could be inferred for this customer: $p_2 \succ_{cust_1} p_1$, $p_2 \succ_{cust_1} p_4$, $p_3 \succ_{cust_1} p_1$, $p_3 \succ_{cust_1} p_4$ (3.2, bottom).

Definition

This allows us to define the function d_s using the observed data s:

$$d_s(\mathbf{c}, x_i, x_j) := \delta\left(s(\mathbf{c}, x_i) > s(\mathbf{c}, x_j)\right) \cdot \left(s(\mathbf{c}, x_i) - s(\mathbf{c}, x_j)\right) \tag{3.19}$$

The δ-term selects only triples, where x_i has been observed more often than x_j. The second term creates one pair for each time x_i is observed more often than x_j.

Set Notation

For a binary s, this can be rewritten in set notation:

$$D_S \subseteq \mathscr{C} \times X_m^2 \tag{3.20}$$

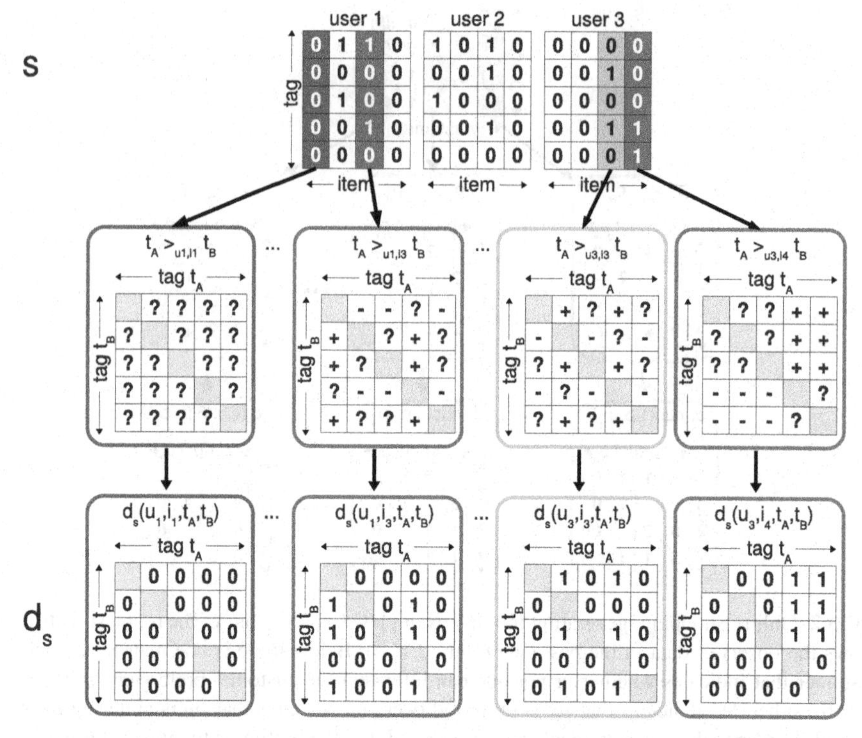

Fig. 3.3 Tag Recommendation: generating training examples d_s for $\succ$ from s. The top shows the observed data s in the past. In the middle are the inferred preferences over tags for each user/ item combination (=context). Note that for context without observation (e.g. (u_1, i_1)), no preferences can be deduced. The bottom contains the training data for $\succ$.

And:

$$D_S := \{(\mathbf{c}, x_i, x_j) \in \mathscr{C} \times X_m^2 : (\mathbf{c}, x_i) \in S \wedge (\mathbf{c}, x_j) \notin S\} \tag{3.21}$$

Example

- **Online shop:** Figure 3.2 shows an example for the online shopping case. Here, the context is the customer. For each customer, the products are compared. If a user has bought a product in the past, we assume that he prefers this product over the products he has not bought in the past. No preference is inferred over product pairs the user has both not bought in the past or product pairs he has both bought in the past.
- **Tag Recommender:** In the case of tag recommender (fig. 3.3), the context contains both the user and the item. That means for every user-item pair a comparison over tags is made. In contrast to the online shop scenario, here most context-pairs (e.g. $\mathbf{c} = (u_1, i_1)$) have no training data: $\forall x \in X_m : s(\mathbf{c}, x) = 0$. In this case,

no training data d_s for ranking is inferred. For context with observations (e.g. $\mathbf{c} = (u_1, i_3)$), the tags with observations within this context are preferred over tags without observations within this context.

3.3.2 Complexity

In the following, we analyze the complexity of d_s in terms of non-zero entries. This complexity is important, as our optimization approach will iterate over these non-zero entries. We start with an analysis with the special case of sets, i.e. S and D_S.

$$\begin{aligned} |D_S| &= |\{(\mathbf{c}, x_i, x_j) \in \mathscr{C} \times X_m^2 : (\mathbf{c}, x_i) \in S \wedge (\mathbf{c}, x_j) \notin S\}| \\ &= \sum_{\mathbf{c} \in \mathscr{C}} |\{(x_i, x_j) \in X_m^2 : (\mathbf{c}, x_i) \in S \wedge (\mathbf{c}, x_j) \notin S\}| \end{aligned} \tag{3.22}$$

Let $S_{\mathbf{c}}^+$ be the set of all instances $x \in X_m$ that are observed and $S_{\mathbf{c}}^-$ be the unobserved ones within the context $\mathbf{c}$:

$$S_{\mathbf{c}}^+ := \{x \in X_m : (\mathbf{c}, x) \in S\} \tag{3.23}$$

$$S_{\mathbf{c}}^- := \{x \in X_m : (\mathbf{c}, x) \notin S\} = X_m \setminus S_{\mathbf{c}}^+ \tag{3.24}$$

This allows to rewrite eq. (3.22):

$$|D_S| = \sum_{\mathbf{c} \in \mathscr{C}} |S_{\mathbf{c}}^+| \, |S_{\mathbf{c}}^-| \tag{3.25}$$

And because the number of positive observations is usually very small compared to all instances, i.e. $|S| \ll |\mathscr{X}|$ and $|S_{\mathbf{c}}^+| \ll |X_m|$ we can approximate the size of $S_{\mathscr{C}}^-$ by the size of X_m:

$$|D_S| < \sum_{\mathbf{c} \in \mathscr{C}} |S_{\mathbf{c}}^+| \, |X_m| = |X_m| \sum_{\mathbf{c} \in \mathscr{C}} |S_{\mathbf{c}}^+| = |X_m| \, |S| \tag{3.26}$$

That means the size of D_S is linear in the number of non-zero observations in S and linear in the size of variables to rank X_m.

Multiple observations

In case of multiple observations, where S is no longer a set but a function s, we can make a similar analysis using sets as an upper bound for non-zero elements in d_s. The quantity of interest is:

$$|d_s| := |\{(\mathbf{c}, x_i, x_j) \in \mathscr{C} \times X_m^2 : d_s(\mathbf{c}, x_i, x_j) > 0\}| \tag{3.27}$$

using the definition of d_s:

$$
\begin{aligned}
|d_s| &= |\{(\mathbf{c},x_i,x_j) \in \mathscr{C} \times X_m^2 : \delta\,(s(\mathbf{c},x_i) > s(\mathbf{c},x_j)) \cdot (s(\mathbf{c},x_i) - s(\mathbf{c},x_j)) > 0\}| \\
&= |\{(\mathbf{c},x_i,x_j) \in \mathscr{C} \times X_m^2 : s(\mathbf{c},x_i) > s(\mathbf{c},x_j)\}| \\
&= \sum_{\mathbf{c}\in\mathscr{C}} |\{(x_i,x_j) \in X_m^2 : s(\mathbf{c},x_i) > s(\mathbf{c},x_j)\}| \qquad (3.28)
\end{aligned}
$$

Now let:

$$S_{\mathbf{c}}^{+} := \{x \in X_m : s(\mathbf{c},x) > 0\} \qquad (3.29)$$

$$S_{\mathbf{c}}^{0} := \{x \in X_m : s(\mathbf{c},x) = 0\} = X_m \setminus S_{\mathbf{c}}^{+} \qquad (3.30)$$

So we can bound the number of non-zero entries to:

$$|d_s| \leq \sum_{\mathbf{c}\in\mathscr{C}} |S_{\mathbf{c}}^{+}||S_{\mathbf{c}}^{0}| \qquad (3.31)$$

And again we can approximate this because in practice $|S_{\mathbf{c}}^{+}| \ll |X_m|$:

$$|d_s| \leq \sum_{\mathbf{c}\in\mathscr{C}} |S_{\mathbf{c}}^{+}||X_m| = |X_m| \sum_{\mathbf{c}\in\mathscr{C}} |S_{\mathbf{c}}^{+}| = |X_m|\,|S| \qquad (3.32)$$

That means again, the number of non-zero entries of d_s is linear both in X_m and the number of observed instances S.

Comparison to Element-wise Training Data

In the next chapter, we will describe how to optimize a model against the non-zero elements in d_s. We have seen, that there are approximately $|S|\,|X_m|$ triples. Now we want to compare this to a method that approximates the observations s directly. In general, for learning a classifier for approximating s, one cannot just train on the non-zero elements but optimization has to be done on $\mathscr{X}$. In contrast when the optimization is performed for ranking using d_s, one can learn only on the non-zero pairs because every positive pair induces also the negative counterpart.

In total, that means that naive reconstruction of s has $|\mathscr{X}|$ training instances whereas learning on non-negative pairs has $|X_m|\,|S|$ training instances. For two mode cases, S is usually larger than $\mathscr{C} = X_1$ and thus the number of training instances for the pair approach is larger than for the reconstruction approach, because $|X_2|\,|S| > |X_1|\,|X_2| = |\mathscr{X}|$. For three or more modes, S is usually much smaller than $|\mathscr{C}| = |X_1|\dots|X_{m-1}|$ and thus the pair approach has less instances than reconstruction of s, i.e. $|X_m|\,|S| \ll |X_1|\dots|X_{m-1}|\,|X_m| = |\mathscr{X}|$.

For enhancing the reconstruction approach one could try to reconstruct only instances within context that has some observations. In this case, the runtime of the element-wise approach would be much smaller than $|\mathscr{X}|$. But with this change, the approach would not reconstruct s any more but it would solve another problem.

3.4 Expressing Rankings by Real Valued Functions

After we have described the task of ranking and how to generate pairwise training data from s, we will next discuss how to reformulate the ranking problem such that it can be modelled efficiently.

Modelling and estimating (strict) total orders on categorical domains (like X_m) is difficult because the order is defined over binary pairs with constraints (eqs. (3.10)-(3.12)). Instead, we propose to map the problem into the real values where a strict total order exists trivially.

Let y be a function from $\mathscr{X}$ to the reals[3]:

$$y : \mathscr{X} \rightarrow \mathbb{R} \tag{3.33}$$

In the following, we will show how y and $\succ$ can be linked. Using y instead of $\succ$ has the advantage that on $\mathbb{R}$ an order exists, that satisfies irreflexivity and transitivity. On the other hand, there is no unique y to represent $\succ$.

3.4.1 *Transformation of Rankings*

First, we will show how to obtain a ranking $\succ^y$ from y. Then, we show that $\succ^y$ satisfies transitivity and irreflexivity but does not have to be connex – i.e. two elements can be placed on the same rank.

Let $y : \mathscr{X} \rightarrow \mathbb{R}$ be a function, then a context-aware ranking $\succ^y$ can be obtained from y by:

$$\forall \mathbf{c} \in \mathscr{C}, x_i, x_j \in X_m : \; x_i \succ^y_{\mathbf{c}} x_j :\Leftrightarrow y(\mathbf{c}, x_i) > y(\mathbf{c}, x_j) \tag{3.34}$$

Lemma 3.1. *For any function y, the ranking $\succ^y$ satisfies transitivity and irreflexivity.*

Proof. First, we show transitivity:

$$\begin{aligned}
&\left(\forall \mathbf{c} \in \mathscr{C},\ \forall x_i, x_j, x_k \in X_m : \; x_i \succ^y_{\mathbf{c}} x_j \wedge x_j \succ^y_{\mathbf{c}} x_k \Rightarrow x_i \succ^y_{\mathbf{c}} x_k\right) \\
\Leftrightarrow &(\forall \mathbf{c} \in \mathscr{C},\ \forall x_i, x_j, x_k \in X_m : \\
&\qquad y(\mathbf{c}, x_i) > y(\mathbf{c}, x_j) \wedge y(\mathbf{c}, x_j) > y(\mathbf{c}, x_k) \Rightarrow y(\mathbf{c}, x_i) > y(\mathbf{c}, x_k))
\end{aligned}$$

This holds for any y because $>$ on $\mathbb{R}$ is transitive.

Secondly for irreflexivity:

$$\begin{aligned}
&\left(\forall \mathbf{c} \in \mathscr{C},\ \forall x_i \in X_m : \; \neg(x_i \succ^y_{\mathbf{c}} x_i)\right) \\
\Leftrightarrow &(\forall \mathbf{c} \in \mathscr{C},\ \forall x_i \in X_m : \; \neg(y(\mathbf{c}, x_i) > y(\mathbf{c}, x_i)))
\end{aligned}$$

As $>$ is irreflexive on $\mathbb{R}$, this always holds.

[3] Note that the target of y is $\mathbb{R}$ and not $\mathbb{N}$ because y is used as a free scoring function and not to estimate rank positions directly.

Remark 3.1. The ranking $\succ^y$ defined by y is not always a strict total order.

Proof. This can be shown by contradiction. First, we define y:

$$y(\mathbf{x}) := 0, \quad \forall \mathbf{x} \in \mathscr{X} \tag{3.35}$$

According to eq. (3.11), the ranking has to be connex:

$$\begin{aligned}
&(\forall \mathbf{c} \in \mathscr{C},\ \forall x_i, x_j \in X_m :\ x_i \succ^y_{\mathbf{c}} x_j \vee x_j \succ^y_{\mathbf{c}} x_j \vee x_i = x_j)\\
\Leftrightarrow &(\forall \mathbf{c} \in \mathscr{C},\ \forall x_i, x_j \in X_m :\ y(\mathbf{c}, x_i) > y(\mathbf{c}, x_j) \vee y(\mathbf{c}, x_j) > y(\mathbf{c}, x_i) \vee x_i = x_j)\\
\Leftrightarrow &(\forall \mathbf{c} \in \mathscr{C},\ \forall x_i, x_j \in X_m :\ (0 > 0) \vee (0 > 0) \vee x_i = x_j)\\
\Leftrightarrow &(\forall \mathbf{c} \in \mathscr{C},\ \forall x_i, x_j \in X_m :\ x_i = x_j)
\end{aligned}$$

This obviously does not hold for any non-trivial domain (i.e. $|X_m| > 1$) and thus the contradiction is shown.

In total, this means that every y defines a context-aware ranking but it might place several variable instances on the 'same' rank. To overcome this, we assume a random order for two instances with the same value y. With this modification $\succ^y$ is a strict total order.

3.4.2 Expressiveness

Next, we discuss the expressiveness of using y to represent $\succ$ for countable[4] X_m. First we show that y can express any $\succ$ and secondly, we show that there are many different y to express the same $\succ$.

Lemma 3.2. *For countable X_m, y can express any ranking $\succ$.*

Proof. Define y:

$$y(\mathbf{c}, x_i) := |\{x_j \in X_m : x_i \succ_{\mathbf{c}} x_j\}| \tag{3.36}$$

Now:

$$\begin{aligned}
\forall \mathbf{c} \in \mathscr{C}, x_i, x_j \in X_m : y(\mathbf{c}, x_i) > y(\mathbf{c}, x_j)&\\
&\Leftrightarrow |\{x_k \in X_m : x_i \succ_{\mathbf{c}} x_k\}| > |\{x_k \in X_m : x_j \succ_{\mathbf{c}} x_k\}|\\
&\Leftrightarrow x_i \succ_{\mathbf{c}} x_j
\end{aligned}$$

Remark 3.2. There is no unique y to express $\succ$.

Proof. It is easy to show this by defining $y' := ay + b$ with $a \in \mathbb{R}^+, b \in \mathbb{R}$. But also any other strictly monotonically increasing function f can be applied. Furthermore

[4] Throughout this work, we always deal with countable domains to rank. Even more, X_m is usually finite.

f can be individual for each $\mathbf{c}$, i.e. it can differ. Let $f_\mathbf{c} : \mathbb{R} \to \mathbb{R}$ be a strictly monotonically increasing function:

$$\forall a,b \in \mathbb{R} : f_\mathbf{c}(a) > f_\mathbf{c}(b) \Leftrightarrow a > b \tag{3.37}$$

We define y' as:

$$\forall \mathbf{c} \in \mathscr{C}, x \in X_m : y'(\mathbf{c},x) := f_\mathbf{c}(y(\mathbf{c},x)) \tag{3.38}$$

Now it is easy to show that both y' and y define the same ranking $\succ$:

$$\begin{aligned} \forall \mathbf{c} \in \mathscr{C}, x_i, x_j \in X_m : x_i \succ_\mathbf{c} x_j &\Leftrightarrow y(\mathbf{c},x_i) > y(\mathbf{c},x_j) \\ &\Leftrightarrow f_\mathbf{c}(y(\mathbf{c},x_i)) > f_\mathbf{c}(y(\mathbf{c},x_j)) \Leftrightarrow y'(\mathbf{c},x_i) > y'(\mathbf{c},x_j) \end{aligned}$$

3.4.3 Discussion

In this section, we have shown how to transform between rankings $\succ$ and real value functions y. We have pointed out that (i) for every ranking $\succ$ there exists many different y and (ii) for every y there exists one ranking $\succ$ – that does not have to be total, i.e. it might place many instances on the same rank. To overcome this, instances on the same rank are ordered randomly. This means a function y that assigns the same value to all instances corresponds to a random ordering.

It is important to note the difference between y and s. One might express y by s, but this would mean that all unobserved instances are placed randomly because for all of them $y(\mathbf{x}) = s(\mathbf{x}) = 0$. Instead, y and s are not related directly with each other. In particular, there is no direct observation of any value y. Instead, d_s defines pairwise observations for learning y – i.e. the (inferred) pair $d_s(\mathbf{c},x_a,x_b) = 1$ defines a relation between $y(\mathbf{c},x_a)$ and $y(\mathbf{c},x_b)$ that is $y(\mathbf{c},x_a) \overset{!}{>} y(\mathbf{c},x_b)$.

3.5 Evaluation Metrics

Finally, we describe common evaluation metrics for ranking problems. We will later use these metrics to assess the empirical quality of our methods. We assume there is a set C of context given for which the evaluation should be made:

$$C = \{\mathbf{c}_1, \mathbf{c}_2, \ldots\} \tag{3.39}$$

For each of these context, the ground truth of variable instances is given as $S_\mathbf{c} \subseteq X_m$. For example, in the online shopping scenario it is known that the customer u (the context) buys a certain product i next; thus $i \in S_u$. Or for tagging, it is known that a user u tags a certain item i (user+item=context) with a set of tags $\{t_A, t_B\}$; that means $S_{u,i} = \{t_A, t_B\}$. Usually, the training data S and the evaluation data is disjoint.

Our evaluation measures are all based on evaluating a ranking. With $\hat{y}$, we can rank the instances of $x \in X_m$ given a context $\mathbf{c}$. Using $\hat{y}$, we can uniquely[5] assign for each instance x an estimated rank $\hat{r}_\mathbf{c}$ in the list sorted by $\hat{y}$:

$$\hat{r}_\mathbf{c} : X_m \rightarrow \{1, \ldots, |X_m|\}, \quad \hat{r}_\mathbf{c} \text{ bijective} \tag{3.40}$$

with:

$$\forall x_i, x_j \in X_m : \hat{r}_\mathbf{c}(x_i) < \hat{r}_\mathbf{c}(x_j) \Leftrightarrow \hat{y}(\mathbf{c}, x_i) > \hat{y}(\mathbf{c}, x_j) \tag{3.41}$$

Note that $\hat{r}_\mathbf{c}$ is a bijective mapping.

Half-life-utility (HLU)

The HLU aka 'Breese score' (Breese et al, 1998) scores the elements with exponential decay:

$$\mathrm{HLU}(S_\mathbf{c}, \hat{r}_\mathbf{c}) := 100 \frac{\sum_{r=1}^{|X_m|} \delta(\hat{r}_\mathbf{c}^{-1}(r) \in S_\mathbf{c}) 2^{-\frac{r-1}{\alpha-1}}}{\sum_{r=1}^{|S_\mathbf{c}|} 2^{-\frac{r-1}{\alpha-1}}} \tag{3.42}$$

The HLU is bound to $[0, 100]$, where 100 is a perfect score.

We will report the average HLU over the test context C:

$$\mathrm{HLU}(C) := \frac{1}{|C|} \sum_{\mathbf{c} \in C} \mathrm{HLU}(S_\mathbf{c}, \hat{r}_\mathbf{c}) \tag{3.43}$$

Precision and recall

Often, a limited set of items is recommended. Precision and recall evaluate such top-N lists. Therefore the first N predictions from the predicted list are taken and compared to $S_\mathbf{c}$. Precision measures the ratio of how many items in the predicted list are in $S_\mathbf{c}$. Whereas recall measures how many of the true items $S_\mathbf{c}$ are covered by the top-N list:

$$\mathrm{Top}(\hat{r}_\mathbf{c}, N) := \{\hat{r}_\mathbf{c}^{-1}(1), \ldots, \hat{r}_\mathbf{c}^{-1}(N)\} \tag{3.44}$$

$$\mathrm{Precision}(S_\mathbf{c}, \hat{r}_\mathbf{c}, N) := \frac{|\mathrm{Top}(\hat{r}_\mathbf{c}, N) \cap S_\mathbf{c}|}{N} \tag{3.45}$$

$$\mathrm{Recall}(S_\mathbf{c}, \hat{r}_\mathbf{c}, N) := \frac{|\mathrm{Top}(\hat{r}_\mathbf{c}, N) \cap S_\mathbf{c}|}{|S_\mathbf{c}|} \tag{3.46}$$

[5] In case of identical values $\hat{y}$ for two instances, we take an arbitrary fixed order within these two instances.

Both precision and recall are bound to the interval $[0,1]$ where the best value is 1. In reality, the precision and recall are also bounded by fixing N. E.g. if N is fixed to 1 and $|S_\mathbf{c}| = 2$, than the maximal achievable recall is $\frac{1}{2}$.

Besides average precision and average recall, we report the f-measure (harmonic mean) over these mean values:

$$\text{Precision}(C,N) := \frac{1}{|C|} \sum_{\mathbf{c} \in C} \text{Precision}(S_\mathbf{c}, \hat{r}_\mathbf{c}, N) \tag{3.47}$$

$$\text{Recall}(C,N) := \frac{1}{|C|} \sum_{\mathbf{c} \in C} \text{Recall}(S_\mathbf{c}, \hat{r}_\mathbf{c}, N) \tag{3.48}$$

$$\text{F-Measure}(C,N) := \frac{2 \cdot \text{Precision}(C,N) \cdot \text{Recall}(C,N)}{\text{Precision}(C,N) + \text{Recall}(C,N)} \tag{3.49}$$

Also for the F-Measure, the best value is 1 and the worst 0.

Area under the ROC curve (AUC)

The AUC measures the pairwise classification rate:

$$\text{AUC}(S_\mathbf{c}, \hat{r}_\mathbf{c}) := \frac{1}{|S_\mathbf{c}| \cdot |X_m \setminus S_\mathbf{c}|} \sum_{x_i \in S_\mathbf{c}} \sum_{x_j \in X_m \setminus S_\mathbf{c}} \delta(\hat{r}_\mathbf{c}(i) < \hat{r}_\mathbf{c}(j)) \tag{3.50}$$

The AUC is bound to $[0,1]$ where 1 is the best value. The AUC of a random order is 0.5.

Again, we report the average AUC over all given context:

$$\text{AUC}(C) := \frac{1}{|C|} \sum_{\mathbf{c} \in C} \text{AUC}(S_\mathbf{c}, \hat{r}_\mathbf{c}) \tag{3.51}$$

Recently, Hand (2009) has shown that the AUC uses different missclassification cost distributions for different classifiers when evaluating on the same problem. Nevertheless, we also use AUC as one of our evaluation measures because it is widely used in the literature and AUC does not depend on an evaluation hyperparameter like N for precision, recall and F-Measure or like α for the HLU.

Other measures

In the information retrieval community two further measures are used to evaluate rankings:

- **Mean average precision (MAP)**
  The average precision (Buckley and Voorhees, 2000) is defined as:

$$\mathrm{AP}(S_{\mathbf{c}}, \hat{r}_{\mathbf{c}}, N) := \frac{1}{|S_{\mathbf{c}}|} \sum_{i=1}^{N} \mathrm{Precision}(S_{\mathbf{c}}, \hat{r}_{\mathbf{c}}, i) \cdot \delta(\hat{r}_{\mathbf{c}}^{-1}(i) \in S_{\mathbf{c}}) \tag{3.52}$$

where N is the list length. The mean average precision (MAP) is the average over all given context:

$$\mathrm{MAP}(C, N) := \frac{1}{|C|} \sum_{\mathbf{c} \in C} \mathrm{MAP}(S_{\mathbf{c}}, \hat{r}_{\mathbf{c}}, N) \tag{3.53}$$

One can set $N = |X_m|$ to generate the average precision at all relevant items.

- **Normalized discounted cumulative gain (NDCG)**
  Like HLU, in NDCG the weight of ranks decreases non-linear. The discounted cumulative gain (DCG) is defined as:

$$\mathrm{DCG}(S_{\mathbf{c}}, \hat{r}_{\mathbf{c}}, N) := \sum_{i=1}^{N} \frac{1}{\log_2(1+i)} \delta(r_{\mathbf{c}}^{-1}(i) \in S_{\mathbf{c}}) \tag{3.54}$$

The NDCG is normalized by the ideal DCG (IDCG), i.e. the DCG of an optimal ranking:

$$\mathrm{NDCG}(S_{\mathbf{c}}, \hat{r}_{\mathbf{c}}, N) := \frac{\mathrm{DCG}(S_{\mathbf{c}}, \hat{r}_{\mathbf{c}}, N)}{\mathrm{IDCG}(S_{\mathbf{c}}, \hat{r}_{\mathbf{c}}, N)} \tag{3.55}$$

where

$$\mathrm{IDCG}(S_{\mathbf{c}}, \hat{r}_{\mathbf{c}}, N) := \sum_{i=1}^{\min(N, |S_{\mathbf{c}}|)} \frac{1}{\log_2(1+i)} \tag{3.56}$$

In our evaluations, we use HLU, F-Measure and AUC. The HLU and NDCG are related because they assign non-linear decreasing weights to ranks and evaluate how much weight the ranking $\hat{r}$ has.

Overlapping Context

In case, a context is evaluated and for this context also training examples exists (i.e. $\exists x \in X_m, \mathbf{c} \in C : s(\mathbf{c}, x) > 0$) we usually do not allow to rerecommend the instances x that are already training examples for this context. Furthermore, we also do not want to evaluate them. That means, instead of dealing with X_m on which rankings should be made and evaluated, we use the subset $X_m^{\mathbf{c}}$ of X_m that does not contain any training examples:

$$X_m^{\mathbf{c}} := \{x \in X_m : s(\mathbf{c}, x) = 0\} \tag{3.57}$$

In such cases where we want to have this restriction, we use $X_m^{\mathbf{c}}$ instead of X_m in the equations (3.40), (3.42), (3.45), (3.46), (3.50), (3.52), (3.54) and (3.56).

References

Breese, J.S., Heckerman, D., Kadie, C.: Empirical analysis of predictive algorithms for collaborative filtering. In: Proceedings of the Fourteenth Conference on Uncertainty in Artificial Intelligence (UAI 1998), pp. 43–52. Morgan Kaufmann, San Francisco (1998)

Buckley, C., Voorhees, E.M.: Evaluating evaluation measure stability. In: SIGIR 2000: Proceedings of the 23rd Annual international ACM SIGIR Conference on Research and Development in Information Retrieval, pp. 33–40. ACM, New York (2000)

Hand, D.J.: Measuring classifier performance: a coherent alternative to the area under the roc curve. Machine Learning 77(1), 103–123 (2009)

Schmidt-Thieme, L.: Compound classification models for recommender systems. In: IEEE International Conference on Data Mining (ICDM 2005), pp. 378–385 (2005)

Chapter 4
Learning Context-Aware Ranking

In this chapter, we propose a learning method for the problem setting of context-aware ranking. This problem setting has been investigated in detail in the last chapter. We have seen, that a context-aware ranking $\succ: \mathscr{C} \times X_m^2$ can be modelled by a real-valued function $y : \mathscr{C} \times X_m \to \mathbb{R}$. Now, we will show how this function can be optimized. The optimization will be done with respect to the pairwise training data d_s, that is inferred from the sparse and incomplete observations s. The whole chapter assumes, that y can be expressed as a differentiable, non-recursive function with a finite set of parameters Θ. This assumption holds for many models, including the factorization models that we will introduce in the next chapter.

First, we propose an optimization criterion based on Bayesian Context-aware Ranking (BCR). The optimization criterion BCR-OPT, is the maximum a posteriori (MAP) estimator of the model parameters Θ. BCR-OPT tries to minimize the classification loss for context-aware ranking on the inferred training data d_s. Furthermore it contains a regularization term that prevents overfitting. Afterwards, we develop the learning algorithm BCR-LEARN which performs the optimization of Θ with respect to BCR-OPT. This algorithm is generic and can be adapted to many model classes including matrix factorization, k-nearest-neighbor, tensor factorization and Markov chains. BCR-LEARN is based on stochastic gradient descent, where the cases are drawn by bootstrapping. This allows the algorithm to converge faster than typical full gradient or context-wise stochastic gradient descent. We show, how samples for BCR-LEARN can be drawn efficiently both in terms of runtime and memory consumption. Finally, we discuss the relationships of BCR-OPT to other criteria like the pairwise approach of AUC optimization or element-wise approaches like weighted regularized least square.

Both the optimization criterion and the learning algorithm that we present here are generic. Thus, this chapter does not discuss any specific model. This will be done in the next chapter, where factorization models are proposed to model y.

4.1 Optimization Criterion (BCR-Opt)

In the last chapter we have shown that learning a ranking $\succ$ can be reformulated as learning a function y. Now, we derive the maximum a posteriori estimator

S. Rendle: Context-Aware Ranking with Factorization Models, SCI 330, pp. 39–50.
springerlink.com

for $\hat{y}$. We assume, that $\hat{y}$ can be fully described by a finite set of parameters Θ – this assumption holds for most methods in machine learning. Thus the estimation of $\hat{y}$ corresponds to estimating Θ.

The MAP estimator for Θ is:

$$\underset{\Theta}{\operatorname{argmax}}\, p(\Theta \mid \succ) = \underset{\Theta}{\operatorname{argmax}}\, p(\succ \mid \Theta)\, p(\Theta) \tag{4.1}$$

Next, we analyse both probabilities $p(\succ \mid \Theta)$ and $p(\Theta)$.

4.1.1 Distribution over Pairs

First, we investigate the probability of each triple $(\mathbf{c}, x_i, x_j) \in \mathscr{C} \times X_m^2$ because these triples define the context-aware ranking $\succ$. Each pair $x_i \succ_{\mathbf{c}} x_j \Leftrightarrow (\mathbf{c}, x_i, x_j) \in \succ$ can be seen as a Bernoulli trial. Thus, we model the probability of a pair given the model parameters by:

$$x_i \succ_{\mathbf{c}} x_j \mid \Theta \;\sim\; \text{Bernoulli}(p_{\mathbf{c},i,j}) \tag{4.2}$$

As the single pairs form a total order, we know:

$$p(x_i \succ_{\mathbf{c}} x_j \mid \Theta) = 1 - p(x_j \succ_{\mathbf{c}} x_i \mid \Theta) \tag{4.3}$$

And also:

$$\delta(x_i \succ_{\mathbf{c}} x_j) = 1 - \delta(x_j \succ_{\mathbf{c}} x_i) \tag{4.4}$$

4.1.2 Distribution over Context-Aware Ranking

The Bernoulli distributions can be used to express $p(\succ \mid \Theta)$. We assume pairwise conditional independence of all $\mathbf{c} \in \mathscr{C}$ given the model parameters Θ. As all variables are discrete, we can write:

$$p(\succ \mid \Theta) = \prod_{(\mathbf{c}, x_i, x_j) \in \mathscr{C} \times X_m^2} p(x_i \succ_{\mathbf{c}} x_j \mid \Theta)^{\delta(x_i \succ_{\mathbf{c}} x_j)} \cdot (1 - p(x_i \succ_{\mathbf{c}} x_j \mid \Theta))^{1-\delta(x_i \succ_{\mathbf{c}} x_j)} \tag{4.5}$$

Due to eqs. (3.10)-(3.12) this simplifies to:

$$\begin{aligned} p(\succ \mid \Theta) &= \prod_{(\mathbf{c}, x_i, x_j) \in \mathscr{C} \times X_m^2} p(x_i \succ_{\mathbf{c}} x_j \mid \Theta)^{\delta(x_i \succ_{\mathbf{c}} x_j)} \cdot p(x_j \succ_{\mathbf{c}} x_i \mid \Theta)^{\delta(x_j \succ_{\mathbf{c}} x_i)} \\ &= \prod_{(\mathbf{c}, x_i, x_j) \in \mathscr{C} \times X_m^2} p(x_i \succ_{\mathbf{c}} x_j \mid \Theta)^{2\,\delta(x_i \succ_{\mathbf{c}} x_j)} \end{aligned} \tag{4.6}$$

4.1.3 Modelling Pairs

Now, we use our model $\hat{y}$ to express $p(x_i \succ_{\mathbf{c}} x_j \mid \Theta)$. Because of eq. (3.34) we can define the ranking $\succ$ given the model parameters Θ using $\hat{y}$:

$$x_i \succ_{\mathbf{c}} x_j | \Theta :\Leftrightarrow \hat{y}(\mathbf{c}, x_i) > \hat{y}(\mathbf{c}, x_j) \Leftrightarrow \hat{y}(\mathbf{c}, x_i) - \hat{y}(\mathbf{c}, x_j) > 0 \tag{4.7}$$

With the logistic function σ, this can be transformed into a probability:

$$p(x_i \succ_{\mathbf{c}} x_j | \Theta) := \sigma(\hat{y}(\mathbf{c}, x_i) - \hat{y}(\mathbf{c}, x_j)) \tag{4.8}$$

4.1.4 Priors on Model Parameters

If the prior distribution of the model parameters is known, $p(\Theta)$ should be modeled by this distribution. Otherwise, a common approach is to assume independent Gaussian priors which leads to the L2-regularizer. We will use this approach in the following.

First, we assume independence of all parameters θ. Secondly, we assume that each θ follows a Normal distribution centered at 0:

$$\theta \sim \mathcal{N}\left(0, \frac{1}{2\lambda_\theta}\right) \tag{4.9}$$

with $\lambda_\theta \in \mathbb{R}^+$.

With these assumption, the prior probability of Θ is:

$$p(\Theta) = \prod_{\theta \in \Theta} \sqrt{\frac{\lambda_\theta}{\pi}} \exp\left(-\lambda_\theta \, \theta^2\right) \tag{4.10}$$

In practice, it is common to treat λ_θ as a hyperparameter which is searched e.g. by a holdout method. Secondly, not for every parameter θ an own hyperparameter λ_θ is searched, but parameters can be grouped and the hyperparameter λ is shared among parameters in the group. A reasonable grouping depends on the model – e.g. for factorization models, the factors of each matrix might share the same regularization parameter λ.

4.1.5 BCR Optimization

These definitions allow to derive the MAP estimator for Θ. As the true ranking $\succ$ is unknown, we use the derived training data d_s (see section 3.3.1). In total, the MAP estimator is:

$$
\begin{aligned}
&\operatorname*{argmax}_{\Theta} p(\Theta \mid \succ) = \operatorname*{argmax}_{\Theta} p(\succ \mid \Theta)\, p(\Theta) \\
&= \operatorname*{argmax}_{\Theta} \prod_{(\mathbf{c},x_i,x_j)\in\mathscr{C}\times X_m^2} \sigma(\hat{y}(\mathbf{c},x_i) - \hat{y}(\mathbf{c},x_j))^{2 d_s(\mathbf{c},x_i,x_j)} \cdot p(\Theta) \\
&= \operatorname*{argmax}_{\Theta} \sum_{(\mathbf{c},x_i,x_j)\in\mathscr{C}\times X_m^2} 2 d_s(\mathbf{c},x_i,x_j) \ln \sigma(\hat{y}(\mathbf{c},x_i) - \hat{y}(\mathbf{c},x_j)) + \ln p(\Theta) \\
&= \operatorname*{argmax}_{\Theta} \sum_{(\mathbf{c},x_i,x_j)\in\mathscr{C}\times X_m^2} d_s(\mathbf{c},x_i,x_j) \ln \sigma(\hat{y}(\mathbf{c},x_i) - \hat{y}(\mathbf{c},x_j)) - \sum_{\theta\in\Theta} \frac{1}{2} \lambda_\theta\, \theta^2 \\
&=: \operatorname*{argmax}_{\Theta} \text{BCR-OPT}
\end{aligned}
\tag{4.11}
$$

BCR can be seen as a weighted regression on ‘pairs’ with regularization, where the weights are d_s. Furthermore, the instances within the pairs overlap in the context.

4.2 Learning Algorithm (BCR-Learn)

In the following, we introduce the learning algorithm BCR-LEARN that optimizes the model parameters Θ for BCR-OPT. This algorithm is based on stochastic gradient descent, i.e. it updates the model parameters for each case instead of computing the full gradient. We show that standard stochastic gradient descent that traverses the data in a sorted way will result in poor convergence. Instead we propose a stochastic gradient descent algorithm using bootstrap sampling from the training data. We show how this sampling can be performed efficiently both in terms of memory and runtime consumption.

4.2.1 Optimization by Gradient Descent

A popular and generic optimization method is gradient descent (for minimization) and gradient ascent (for maximization). The idea of gradient descent methods is to start with an initial guess of the parameters and then iteratively follow the gradients of the objective criterion. In each iteration, the gradient of the objective criterion with respect to the current model parameters is calculated and then a small step into this direction is taken. If the step size is small enough and the problem convex, gradient descent is guaranteed to reach the global optimum. In non-convex problems a local optimum is reached that is not necessarily a global optimum.

The gradient of BCR-OPT with respect to each parameter θ is:

$$
\frac{\partial}{\partial\theta} \text{BCR-OPT} = \sum_{(\mathbf{c},x_i,x_j)\in\mathscr{C}\times X_m^2} \delta_{\mathbf{c},x_i,x_j} \frac{\partial}{\partial\theta} (\hat{y}(\mathbf{c},x_i) - \hat{y}(\mathbf{c},x_j)) - \lambda_\theta\, \theta \tag{4.12}
$$

with

$$
\delta_{\mathbf{c},x_i,x_j} := d_s(\mathbf{c},x_i,x_j)\,(1 - \sigma(\hat{y}(\mathbf{c},x_i) - \hat{y}(\mathbf{c},x_j))) \tag{4.13}
$$

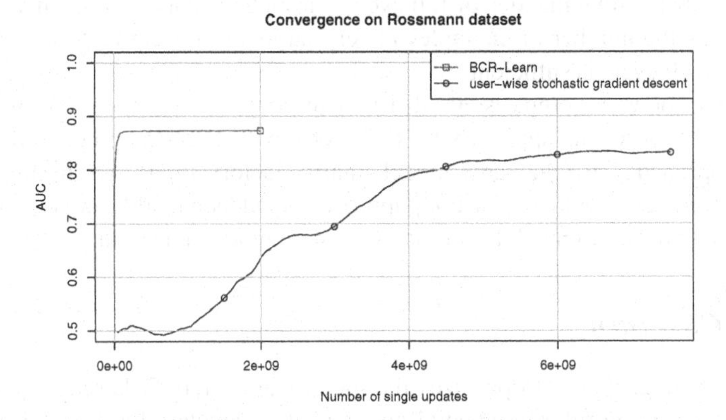

Fig. 4.1 Empirical comparison of the convergence of typical context-wise stochastic gradient descent to our BCR-LEARN algorithm with bootstrap sampling. This example is from a recommender system, where the context is personalization (user) and items should be recommended (see chapter 6).

This full gradient approach leads to a descent in the 'correct' direction, but in our case convergence is slow. As we have $|d_s| \approx |S| \cdot |X_m|$ non-zero training triples (i.e. $d_s(\mathbf{c},x_i,x_j) > 0$), computing the full gradient in each update step is not feasible for mid to large scale problems. Furthermore, for optimizing BCR-OPT with full gradient descent also the skewness in the training pairs leads to poor convergence. Imagine one instance $x_i \in X_m$ that is often positive. Then we have many terms of the form $\hat{y}_{\mathbf{c},x_i}$ in the loss because for many different context $\mathbf{c}$, x_i is compared against all negative instances $x_j \in X_m$ (the dominating class). Thus the gradient for model parameters depending on x_i would dominate largely the gradient. That means very small learning rates would have to be chosen. Secondly, regularization is difficult as the gradients differ much.

The other popular approach is stochastic gradient descent. In this case, for each triple $(\mathbf{c},x_i,x_j) \in \mathscr{C} \times X_m^2$ an update is performed. The gradient for each parameter θ given this triple is:

$$\frac{\partial}{\partial\theta}\text{BCR-OPT} = \delta_{\mathbf{c},x_i,x_j} \frac{\partial}{\partial\theta}(\hat{y}(\mathbf{c},x_i) - \hat{y}(\mathbf{c},x_j)) - \lambda_\theta\,\theta \tag{4.14}$$

In general, this is a good approach for our skew problem but the order in which the training pairs are traversed is crucial. A typical approach that traverses the data in any sorted way (e.g. by context) will lead to poor convergence as there are so many consecutive updates on the same pairs $(\mathbf{c},x_i)$ – i.e. for one pair $(\mathbf{c},x_i) \in S$ there are many $x_j \in X_m$ with $d_s(\mathbf{c},x_i,x_j) > 0$.

To solve this issue we suggest to use a stochastic gradient descent algorithm that chooses the triples randomly (uniformly distributed). With this approach the chances to pick the same pair $(\mathbf{c},x_i)$ in consecutive update steps is small. We suggest to use a bootstrap sampling approach with replacement because stopping can be performed at any step. Abandoning the idea of full cycles through the data is especially useful in our case as the number of examples is very large and for convergence often a fraction of a full cycle is sufficient.

Figure 4.1 shows a comparison[1] of a typical context-wise stochastic gradient descent algorithm to our approach BCR-LEARN with bootstrapping. Both optimization approaches use the same model (matrix factorization) and the hyperparameters have been optimized for both approaches independently. As you can see BCR-LEARN converges much faster than user-wise gradient descent.

4.2.2 BCR-Learn

BCR-LEARN (algorithm 1) optimizes the model parameters Θ based on the observations s. The algorithm is general and can fit parameters for many different models. In the chapters 6, 7 and 8, we will see examples for matrix factorization, k-nearest-neighbor, tensor factorization and Markov chain models. In the first step of the algorithm, the model parameters are initialized with a first guess. In general, no information about the model parameters is available a priori and thus they are initialized with random numbers, e.g. drawn from a normal distribution. Then, the model parameters are iteratively fitted.

In each iteration one case $(\mathbf{c}, x_i, x_j)$ is drawn and the gradient (eq. (4.14)) is computed. The δ-term can be seen as a parameter-independent weight for the update: the larger the error, the larger the update. Then each parameter that is related to the case is updated. Mostly only a small fraction of parameters has to be updated – e.g. in factorization models, only the factors of the variable instances within the case have a gradient larger than 0. Examples for this are shown in chapter 6, 7 and 8. The gradient of $\hat{y}$ depends on the model itself and has to be derived for each model class individually.

The iterative fitting procedure is continued until a stopping criterion is reached. A common approach for gradient descent algorithms is to iterate until the loss on the training data converges. But in our case, we stop after a fixed number of iterations. This number of iterations is chosen on a holdout set. An advantage of this approach is that this 'early stopping' (before convergence of the loss) can help to prevent overfitting. Thus it can be seen as an additional regularizer.

4.2.3 Drawing of Training Cases

In each iteration of BCR-LEARN, a triple $(\mathbf{c}, x_i, x_j)$ is drawn uniformly from $\mathscr{C} \times X_m^2$. A straight-forward implementation would be to draw $x_1 \in X_1, \ldots, x_{m-1} \in X_{m-1}$ and $x_i, x_j \in X_m$ independently. With this implementation, the probability for each triple $(\mathbf{c}, x_i, x_j)$ is equal, i.e. we would have a uniform draw. On the other hand, we know that the training pairs are very sparse ($|d_s| \ll |\mathscr{C} \times X_m^2|$) – see eq. (3.32). Thus for a random draw it is very likely that $d_s(\mathbf{c}, x_i, x_j) = 0$. This means that this triple is useless for making any informative update on the parameters.

[1] Details about the dataset and evaluation method can be found in section 6.6.

Algorithm 1 BCR-LEARN

Input: training data s, learning rate α, regularization parameters λ_θ
Output: model parameters Θ
1: initialize Θ from $\mathcal{N}(0,\sigma^2)$
2: **repeat**
3: draw $(\mathbf{c},x_i,x_j)$ uniformly from $\mathscr{C} \times X_m^2$
4: $\delta_{\mathbf{c},i,j} \leftarrow d_s(\mathbf{c},x_i,x_j)\,(1-\sigma(\hat{y}(\mathbf{c},x_i)-\hat{y}(\mathbf{c},x_j)))$
5: **for** $\theta \in \Theta$ **do**
6: $\theta \leftarrow \theta + \alpha\left(\delta_{\mathbf{c},i,j}\,\frac{\partial}{\partial\theta}(\hat{y}(\mathbf{c},x_i)-\hat{y}(\mathbf{c},x_j)) - \lambda_\theta\,\theta\right)$
7: **end for**
8: **until** convergence
9: **return** Θ

A naive solution would be to enumerate all triples $(\mathbf{c},x_i,x_j)$ with $d_s(\mathbf{c},x_i,x_j) > 0$ and then draw from this set uniformly. Even though this would guarantee that only non-zero triples are drawn, enumerating all such triples is not feasible because the number of these triples is large ($|d_s| \approx |S|\,|X_m|$).

Instead, we can change the drawing scheme such that only triples are drawn that are likely to have $d_s(\mathbf{c},x_i,x_j) > 0$. Analyzing the definition of d_s (eq. 3.19), we see that $d_s(\mathbf{c},x_i,x_j) > 0$ only if $s(\mathbf{c},x_i) > 0$. Thus, we can formulate an alternative drawing scheme for triples that is based on rejection sampling:

1. repeat
   a. draw $(\mathbf{c},x_i)$ uniformly from S (see eq. (3.5))
   b. draw x_j uniformly from X_m
2. until $d_s(\mathbf{c},x_i,x_j) > 0$

This drawing scheme has two advantages: (1) there is no additional overhead for storing triples, because the procedure works directly with the observations S. And (2) it is very likely to find cases that are positive, such that a redraw is usually not necessary. A redraw is only necessary if $s(\mathbf{c},x_j) \geq s(\mathbf{c},x_i)$. Usually within a context $\mathbf{c}$ the set of observed instances is very small ($|\{x \in X : s(\mathbf{c},x) > 0\}| \ll |X_m|$), thus it is very unlikely to randomly select $x_j \in X_m$ that are observed (non-zero).

For cases with binary s, the proposed algorithm corresponds to drawing a positiv case $(\mathbf{c},x_i) \in S$ and a negative one $(\mathbf{c},x_j) \notin S$.

4.3 Alternative Optimization Criteria

Next, we compare the BCR optimization criterion to other optimization approaches. We start with other pairwise losses and show the relation of BCR-OPT to AUC optimization. Then, we discuss several element-wise losses that try to reconstruct the training data s.

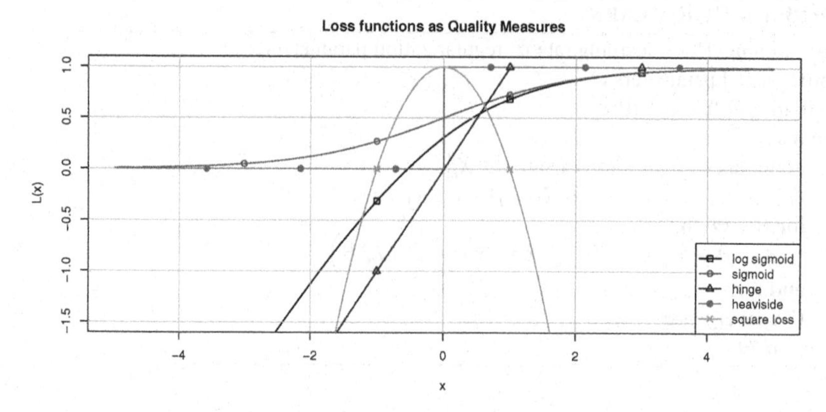

Fig. 4.2 Comparison of loss functions. The heaviside-function is often approximated by the sigmoid σ. For BCR, the MAP derivation suggests to use $\ln \sigma$ as loss functions.

4.3.1 Pairwise Losses

BCR-OPT can be seen as minimizing the error of ranking pairs within d_s.

$$\underset{\Theta}{\operatorname{argmax}} \sum_{(\mathbf{c},x_i,x_j)\in\mathscr{C}\times X_m^2} d_s(\mathbf{c},x_i,x_j)\, L(\hat{y}(\mathbf{c},x_i) - \hat{y}(\mathbf{c},x_j)) - \sum_{\theta\in\Theta} \frac{1}{2}\lambda_\theta\, \theta^2 \tag{4.15}$$

where the loss L is the log-sigmoid:

$$L(x) = \ln \sigma(x) \tag{4.16}$$

This loss is justified by the Bayesian analysis where the MAP estimator leads to this loss when the probabilities are modelled with σ. In general, also other losses within BCR-OPT are possible. A graphical comparison of loss functions[2] is shown in figure 4.2.

Hinge-Loss

The hinge loss is used for example in maximum-margin classifiers like SVMs or maximum margin matrix factorization (Srebro et al, 2005). The hinge loss is usually defined as:

$$L^*(x) := \max(0, 1-x) \tag{4.17}$$

When using the loss as a quality measure like in BCR, we can change it to:

$$L(x) = 1 - L^*(x) = 1 + \min(0, -1+x) = \min(1,x) \tag{4.18}$$

[2] Note that all 'losses' are transformed to quality measures, because BCR-OPT maximizes the loss. Losses like square-loss and the hinge loss are typically used in minimizers.

Analogies to Area-under the ROC curve optimization

Comparing the AUC quality measure (eq. (3.50)) with BCR-OPT (eq. (4.11)), it is easy to grasp the analogies. First, we reformulate the AUC for a context $\mathbf{c}$ as:

$$\mathrm{AUC}(S_{\mathbf{c}}) := \frac{1}{|S_{\mathbf{c}}| \cdot |X_m \setminus S_{\mathbf{c}}|} \sum_{x_i \in S_{\mathbf{c}}} \sum_{x_j \in X_m \setminus S_{\mathbf{c}}} \delta(\hat{y}(\mathbf{c}, x_i) > \hat{y}(\mathbf{c}, x_j)) \tag{4.19}$$

where $S_{\mathbf{c}} = \{x_i \in X_m : (\mathbf{c}, x_i) \in S\}$[3]. And the AUC over all context is:

$$\mathrm{AUC} := \frac{1}{|\mathscr{C}|} \sum_{\mathbf{c} \in \mathscr{C}} \frac{1}{|S_{\mathbf{c}}| \cdot |X_m \setminus S_{\mathbf{c}}|} \sum_{x_i \in S_{\mathbf{c}}} \sum_{x_j \in X_m \setminus S_{\mathbf{c}}} \delta(\hat{y}(\mathbf{c}, x_i) > \hat{y}(\mathbf{c}, x_j)) \tag{4.20}$$

With the observations D_S, this can be rewritten as:

$$\mathrm{AUC} := \sum_{(\mathbf{c}, x_i, x_j) \in D_S} z_{\mathbf{c}} \, \delta(\hat{y}(\mathbf{c}, x_i) > \hat{y}(\mathbf{c}, x_j)) \tag{4.21}$$

where $z_{\mathbf{c}}$ is a normalizing constant that assures that all context is given the same weight.

Now besides the normalization constants $z_{\mathbf{c}}$, AUC and BCR-OPT differ only in the loss function. For AUC, the loss is usually written as the Heaviside function H, which can be expressed by the delta function:

$$L(x) := \delta(x > 0) = H(x) := \begin{cases} 1, & x > 0 \\ 0, & \text{else} \end{cases} \tag{4.22}$$

Instead we use the differentiable loss $\ln \sigma(x)$. Replacing the non-differentiable Heaviside function is common practice when optimizing for AUC (Herschtal and Raskutti, 2004). Often the choice of the substitution is heuristic and a similarly shaped function like σ is used for smoothing H (see figure 4.2). Our derivation of BCR-OPT suggest the alternative substitution of H by $\ln \sigma(x)$ that is motivated by the MLE.

Square-loss

The square-loss is typically used rather for regression than for classification. For the pairwise approach the square loss has to be shifted to the positive side, such that positive differences are enforced. E.g.:

$$L(x) := 1 - (x - 1)^2 \tag{4.23}$$

[3] Note that the AUC is only defined for binary classes. Thus we assume here the training data is set data S.

But the drawback of square-loss in all classification tasks is that it penalizes correct decisions with large values. E.g. in our case if the training case $(\mathbf{c}, x_i, x_j)$ is correctly classified as positive, square-loss would penalize this when the difference of both values are large, i.e. $y(\mathbf{c}, x_i) \gg y(\mathbf{c}, x_j)$.

4.3.2 *Element-Wise Losses*

Next, we will discuss approaches that are not optimizing rankings but the reconstruction of s.

Several work suggests to minimize the least-square error on s for learning factorization models. This is usually motivated by the popular singular value decomposition (SVD), that minimizes square loss. For example Symeonidis et al (2008) use a higher-order SVD for tag recommendation or Hu et al (2008) use a weighted regularized least-square approach for item recommendation. Pan and Scholz (2009) use also a weighted regularized approach for item recommendation, where besides least-square also the hinge-loss is proposed.

Dense Optimization

All approaches listed above, optimize on all elements of $\mathcal{X}$:

$$\operatorname*{argmin}_{\Theta} \sum_{(\mathbf{c}, x_i) \in \mathcal{X}} L(\hat{y}_{\mathbf{c}, x_i}, \delta(s(\mathbf{c}, x_i) > 0)) \tag{4.24}$$

where L is a loss like square loss or hinge loss. That means on large data sets or higher modes, the number of cases is huge. For sparse problems (i.e. many 0 values), there exists fast solvers for square loss problems. For higher-order SVD, Lathauwer et al (2000) proposed an approximation based on unfolding the tensor and solving for each mode a sparse two-mode SVD. Also fast least-square solver with regularization for sparse two-mode settings are available (Hu et al, 2008). Pan and Scholz (2009) extend the square-loss approach to hinge-loss but because similar scaling approaches as for square-loss cannot be applied, they use subsampling and ensemble several estimators.

But besides that the problem can be solved for least-square loss, all the issues mentioned in section 3.1.4 are not addressed: (1) non-observed elements are not necessarily negative, in contrast, the recommender has to choose among them in the future; (2) the negative cases hugely dominate the observed ones and (3) optimization is done for reconstruction, not for ranking.

Weighted Optimization

To overcome some of these problems, both Hu et al (2008) and Pan et al (2008) suggest to weight each element of $\mathcal{X}$. Extended to the general context-aware setting, this can be written as:

$$\underset{\Theta}{\operatorname{argmin}} \sum_{(\mathbf{c},x_i)\in\mathscr{X}} w_{\mathbf{c},x_i} L(\hat{y}_{\mathbf{c},x_i}, \delta(s(\mathbf{c},x_i) > 0)) \tag{4.25}$$

where $w_{\mathbf{c},x_i}$ is a predefined weight. Hu et al (2008) set the weight using the counts s:

$$w_{\mathbf{c},x_i} = 1 + \alpha\, s(\mathbf{c},x_i) \tag{4.26}$$

where α is a global hyperparameter. Pan et al (2008) suggest to set $w_{\mathbf{c},x_i}$ as a global hyperparameter or as a user / item specific constant. The weighting approach might solve the problem of imbalance as positive examples can be weighted higher than negative ones.

Sparse Optimization

In addition to the ideas of Hu et al (2008) and Pan et al (2008), one could also use the weights to make a sparse optimization, that only optimizes within context that has some observations. This would prevent the element-wise approaches to fit completely unobserved context to zero:

$$w_{\mathbf{c},x_i} = \begin{cases} 1, & \text{if } \exists x_j \in X_m : s(\mathbf{c},x_j) > 0 \\ 0, & \text{else} \end{cases} \tag{4.27}$$

This might be extremely useful for settings with a larger number of modes ($m > 2$), because (1) fitting only elements within observed context might improve quality and (2) it reduces the number of training data. In total, this corresponds to the following sparse optimization:

$$\underset{\Theta}{\operatorname{argmin}} \sum_{\mathbf{c}\in\mathscr{C}:\exists x_j\in X_m, s(\mathbf{c},x_j)>0} \sum_{x_i\in X_m} L(\hat{y}_{\mathbf{c},x_i}, \delta(s(\mathbf{c},x_i) > 0)) \tag{4.28}$$

Outlook

In chapter 6 and 7, we will compare element-wise optimization approaches to pairwise ranking approaches. For item recommendation, we will compare a matrix factorization learned with BCR to learning with regularized weighted least-square and standard least-square (SVD). For tag recommendation, we compare a BCR learned tensor model to higher-order SVD and AUC optimization.

References

Herschtal, A., Raskutti, B.: Optimising area under the roc curve using gradient descent. In: ICML 2004: Proceedings of the Twenty-First International Conference on Machine Learning, p. 49. ACM, New York (2004)

Hu, Y., Koren, Y., Volinsky, C.: Collaborative filtering for implicit feedback datasets. In: IEEE International Conference on Data Mining (ICDM 2008), pp 263–272(2008)

Lathauwer, L.D., Moor, B.D., Vandewalle, J.: A multilinear singular value decomposition. SIAM J. Matrix Anal. Appl. 21(4), 1253–1278 (2000)

Pan, R., Scholz, M.: Mind the gaps: weighting the unknown in large-scale one-class collaborative filtering. In: KDD 2009: Proceedings of the 15th ACM SIGKDD International Conference on Knowledge Discovery and Data Mining, pp. 667–676. ACM, New York (2009)

Pan, R., Zhou, Y., Cao, B., Liu, N.N., Lukose, R.M., Scholz, M., Yang, Q.: One-class collaborative filtering. In: IEEE International Conference on Data Mining (ICDM 2008), pp. 502–511 (2008)

Srebro, N., Rennie, J.D.M., Jaakola, T.S.: Maximum-margin matrix factorization. In: Advances in Neural Information Processing Systems, vol. 17, pp. 1329–1336. MIT Press, Cambridge (2005)

Symeonidis, P., Nanopoulos, A., Manolopoulos, Y.: Tag recommendations based on tensor dimensionality reduction. In: RecSys 2008: Proceedings of the 2008 ACM Conference on Recommender Systems, pp. 43–50. ACM, New York (2008)

Chapter 5
Factorization Models

In the last chapters, it was shown that a context-aware ranking $\succ$ can be expressed by a function $y : \mathscr{X} \rightarrow \mathbb{R}$ or equivalently by a tensor $Y \in \mathbb{R}^{\mathscr{X}}$ in case of finite categorical domains X_i. Estimating the full parametrized tensor Y is infeasible because (1) for real-world problems, the number of parameters (i.e. $|\mathscr{X}|$) would be too large – e.g. for the Netflix[1] problem we would need billions of parameters – and (2) even more important, that the observations are typically very sparse which results in poor estimates without any generalization capabilities.

In this chapter, we discuss how to model the tensor Y by a factorization model $\hat{Y}$. In factorization models, each variable instance x_i is expressed by a real valued vector $\mathbf{v} \in \mathbb{R}^k$ of k factors. For reconstructing the tensor $\hat{Y}$, the factors $\mathbf{v}_1, \ldots \mathbf{v}_m$ of each entry $\mathbf{x} = (x_1, \ldots, x_m) \in \mathscr{X}$ are combined. This combination is defined by the factorization model and consists usually of summations and multiplications.

We discuss linear models that are based on the Tucker decomposition which is a tensor product of factorization matrices of each mode. The structure of the Tucker decomposition is defined by its core tensor. This core tensor makes the reconstruction runtime of the Tucker decomposition exponential in the number of modes. By choosing a special core tensor, the Tucker model simplifies to PARAFAC (parallel factor analysis) which has only linear runtime. Furthermore, we will introduce the more specialized model PITF (pairwise interaction tensor factorization) that is a special case of PARAFAC. PITF models all pairwise interactions between variables explicitly. Even though the number pairwise interactions has quadratic growth in the number of modes, we show that for ranking problems the number is linear. We finish our analysis with a comparison of the runtime and expressiveness of these models.

In this whole chapter, we assume that the domain of each X_i is finite. This holds for most domains of our applications like users, items, tags, articles, web pages, words, etc. But obviously this does not hold for continuous domains like time. In chapter 8 and 9, we develop two methods for integrating time into factorization models. The first one is based on a Markov chain and the second one of modelling time variance within each factor.

[1] `http://www.netflixprize.com/`

S. Rendle: Context-Aware Ranking with Factorization Models, SCI 330, pp. 51–65.
springerlink.com

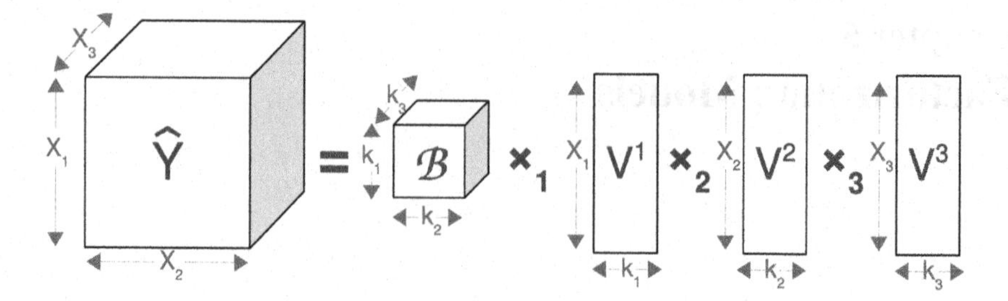

Fig. 5.1 Tucker decomposition (3-mode): The target Y is approximated by a factorization $\hat{Y}$ using a core tensor $\mathscr{B}$ and for each domain X_i one factor matrix V_i of dimensionality k_i.

5.1 Tucker Decomposition (TD)

Tucker decomposition (Tucker, 1966) factorizes a higher-order tensor into a smaller core tensor and one factor matrix for each mode. Figure 5.1 shows an example for three modes ($m = 3$).

The Tucker decomposition (TD) is defined by the following tensor product:

$$\hat{Y}^{\mathrm{TD}} := \mathscr{B} \times_1 V^1 \ldots \times_m V^m \tag{5.1}$$

with model parameters:

$$\begin{aligned} &\mathscr{B} \in \mathbb{R}^{k_1 \times \ldots \times k_m} \\ &V^i \in \mathbb{R}^{|X_i| \times k_i}, \quad \forall i \in \{1, \ldots, m\} \end{aligned} \tag{5.2}$$

$\mathscr{B}$ is called the *core tensor*. V_i is the factor matrix for the categorical variable over the domain X_i where each row $\mathbf{v}^i_j$ describes one variable instance $x_j \in X_i$. Each entry $v^i_{j,f}$ in the matrix V^i is called a factor. k_i is the dimensionality of the factorization for X_i. That means each variable instance is expressed by k_i real numbers.

It is important to note that none of the factors are observed but they are estimated from the data. The number of factors k_i that should be used for describing a variable, is a hyperparameter that is chosen e.g. by the holdout method.

5.1.1 Model Equation

The TD corresponds to the following model equation:

$$\hat{y}^{\mathrm{TD}}_{x_1,\ldots,x_m} := \sum_{f_1=1}^{k_1} \cdots \sum_{f_m=1}^{k_m} b_{f_1,\ldots,f_m} \prod_{i=1}^{m} v^i_{x_i,f_i} \tag{5.3}$$

As we have discussed before, the major problem in context-aware ranking tasks is the sparsity of the data. Factorization models like TD solve this problem by parametrizing variable instances instead of relation instances. The model equation predicts relation instances by combining the individual parametrization of variable instances. This way, the model can generalize to relation instances (combinations of variable instances) that have never been observed jointly. During the learning stage, the parameters for each variable instance are optimized such that using them in the model equation (5.3) leads to an optimal prediction (in our case ranking) for the observed cases (in our case ranking pairs).

5.1.2 *Gradients*

Learning of the model parameters can be done by gradient descent based algorithms like BCR-LEARN (see section 4.2.2). To apply these algorithms, the partial derivatives of the model equation have to be known. The gradients of eq. (5.3) for each model parameter with respect to an instance $\mathbf{x} = (x_1, \ldots, x_m)$ are:

$$\frac{\partial \hat{y}^{\mathrm{TD}}_{x_1,\ldots,x_m}}{\partial b_{f_1,\ldots,f_m}} = \prod_{i=1}^{m} v^i_{x_i,f_i} \tag{5.4}$$

$$\frac{\partial \hat{y}^{\mathrm{TD}}_{x_1,\ldots,x_m}}{\partial v^j_{x_j,f_j}} = \sum_{f_1=1}^{k_1} \cdots \sum_{f_{j-1}=1}^{k_{j-1}} \sum_{f_{j+1}=1}^{k_{j+1}} \cdots \sum_{f_m=1}^{k_m} b_{f_1,\ldots,f_m} \prod_{i=1, i\neq j}^{m} v^i_{x_i,f_i} \tag{5.5}$$

When $\mathbf{x}$ is given, the gradient of each parameter b of the core $\mathscr{B}$ is defined because every factor of the core is always used in the model equation. In contrast to this, the gradients of the factorization matrices are only defined (non zero) for the m rows of the m variable instances in $\mathbf{x}$. All other entries of the factor matrices are not updated when $\mathbf{x}$ is fixed.

5.1.3 *Complexity*

Next, we analyse the complexity of TD in terms of number of parameters, number of operations in the model equation and number of operations for updating with respect to one instance $\mathbf{x}$. An overview of this analysis for all models can be found in table 5.1.

- **Number of free parameters:** The number of free parameters of eq. (5.2) is:

$$\sum_{i=1}^{m} k_i |X_i| + \prod_{i=1}^{m} k_i \tag{5.6}$$

  The first term is the number of parameters for the factor matrices and the second term is the number of core factors. To facilitate the analysis and for better comparability, we assume that all factorization dimensions are equally large, i.e. $k_1 = \ldots = k_m =: k$. Now the number of free parameters is:

$$k \sum_{i=1}^{m} |X_i| + k^m \tag{5.7}$$

As you can see, the number of parameters is exponential in m and polynomial in k.

- **Computation of Model Equation:** Predicting one entry $\mathbf{x}$ of Y with eq. (5.3) requires computing the m nested summations. In total, the number of operations is:

$$(k_1 - 1) \cdot \ldots \cdot (k_m - 1) \cdot m \tag{5.8}$$

With the assumption of equal sized factor dimensions this is:

$$m\,(k-1)^m \in O(mk^m) \tag{5.9}$$

Thus, prediction is exponential in the number of modes m and polynomial in the number of factors.

- **Computation of the Gradients:** Also the computation complexity of the gradient for each parameter is in the same class. There are k^m core factors that are updated; calculating each gradient requires $(m-1)$ operations. Additionally, there are km factors in the matrices that require each $(m-1)\,(k-1)^m$ operations. In total, this leads to:

$$(m^2 - 1)k^m \in O(m^2 k^m) \tag{5.10}$$

Storing intermediate gradients for each factor within a mode reduces this to $O(mk^m)$.

According to this analysis, TD has problems to scale to problems with many modes or large factor dimensions because the number of model parameters as well as prediction and update is exponential in m and polynomial in k.

5.1.4 *Two-Mode Tucker Decomposition*

Next, we want to emphasise that TD for two mode problems is different from typical matrix factorization (Srebro et al, 2005) where the model equation is defined as:

$$\begin{aligned} \hat{y}^{\mathrm{MF}}_{x_1,x_2} &:= \sum_f^k v^1_{x_1,f} \cdot v^2_{x_2,f} \\ &= \langle \mathbf{v}^1_{x_1}, \mathbf{v}^2_{x_2} \rangle \end{aligned} \tag{5.11}$$

The differences between TD and MF can be seen by writing the TD model equation explicitly for $m = 2$:

$$\hat{y}^{\mathrm{TD}}_{x_1,x_2} := \sum_{f_1=1}^{k_1} \sum_{f_2=1}^{k_2} b_{f_1,f_2} \cdot v^1_{x_1,f_1} \cdot v^2_{x_2,f_2} \tag{5.12}$$

Now, the differences are obvious: (1) 2-mode TD contains two nested sums and a full parametrized core matrix. By setting the core matrix to a diagonal matrix, the usual two-mode SVD-scheme $U\Sigma V$ is obtained, where $U = V^1$, $B = \Sigma$ and $V = V^2$.

5.1.5 Higher-Order SVD

Higher-order singular value decomposition (HOSVD) uses the same model equation as TD but HOSVD implies also the choice of the optimization criterion that is mean square error. Thus, HOSVD reconstructs a fully observed tensor Y for minimal square loss. In our case with many missing values, the non-observed values have to be imputed (e.g. by 0) which results in very huge sparsity (see section 3.1.4). We will compare our approaches empirically to HOSVD and SVD in chapter 6 and 7.

5.2 Parallel Factor Analysis (PARAFAC)

The PARAFAC model (Harshman, 1970) is a special case of the Tucker decomposition that assumes a diagonal and constant core tensor. An example for three modes ($m = 3$) is shown in figure 5.2.

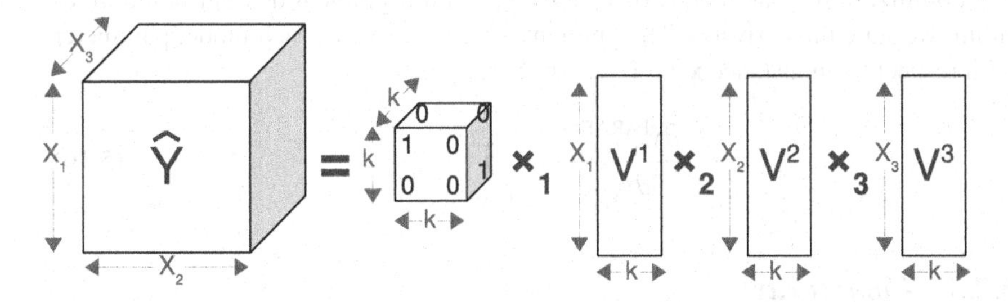

Fig. 5.2 PARAFAC: The model has a diagonal core tensor and the factorization dimensionality is equal for all modes.

PARAFAC uses the same decomposition (eq. (5.1)) and parametrization (eq. (5.2)) as TD but in PARAFAC, the dimensionalities of the factor matrices are identical:

$$\forall i \in \{1,\ldots,m\}: \quad k_i := k \tag{5.13}$$

and the core tensor is diagonal:

$$b_{f_1,\ldots,f_m} := \begin{cases} 1, & \text{if } f_1 = \ldots = f_m \\ 0, & \text{else} \end{cases} \tag{5.14}$$

This leads to model equation where only the factors within the same dimension are linked. This can be seen as a hard-wired core tensor that predefines the interactions between factors whereas in TD this core structure is flexible.

5.2.1 Model Equation

In total, the model equation for PARAFAC is:

$$\hat{y}^{\text{PARAFAC}}_{x_1,\ldots,x_m} := \sum_{f=1}^{k} \prod_{i=1}^{m} v^i_{x_i,f} \tag{5.15}$$

Thus, in PARAFAC all factor values within a given dimension f are multiplied. These products are built over all dimensions and summed up. Even though, the factor matrices itself are not restricted, the model equation is simplified a lot by keeping the core tensor diagonal and constant. Thus, in PARAFAC only the factor matrices have to be learned.

5.2.2 Gradients

For optimizing the parameters of PARAFAC with a gradient descent based algorithm, we state the derivates. The gradients of eq. (5.15) for each model parameter with respect to an instance $\mathbf{x} = (x_1,\ldots,x_m)$ are:

$$\frac{\partial \hat{y}^{\text{PARAFAC}}_{x_1,\ldots,x_m}}{\partial v^j_{x_j,f}} = \prod_{i=1,i\neq j}^{m} v^i_{x_i,f_i} \tag{5.16}$$

5.2.3 Complexity

Next, we will show the influence of fixing the core on the number of free parameters and the computation of prediction and gradients.

- **Number of free parameters:** In general, PARAFAC uses the same decomposition as TD, but by fixing the core tensor to a constant tensor eq. (5.14), the number of free parameters is reduced to:

$$k \sum_{i=1}^{m} |X_i| \tag{5.17}$$

  Now, the number of parameters is linear in k and the sizes of the domains.
- **Computation of Model Equation:** Predicting one entry $\mathbf{x}$ of Y with PARAFAC (eq. (5.15)) requires summation over k products of size m. In total, the number of operations is:

$$(k-1)\,(m-1) \in O(km) \tag{5.18}$$

  Thus, the computation of one element is both linear in k and m.
- **Computation of the Gradients:** For updating with respect to a fixed entry $\mathbf{x}$, all factors of the related variable instances are updated. In total, the number of

updated factors is $k \cdot m$. And for each update with eq. (5.16), the number of operations to calculate the gradient is $m-2$. This results in a total number of update operations of:

$$k \cdot m \cdot (m-2) \in O(km^2) \tag{5.19}$$

By precalculating $\alpha_f := \prod_{i=1}^{m} v^i_{x_i,f}$ for each factor, the calculation of the gradient with respect to $v^i_{x_i,f}$ can be performed in $O(1)$ with $\frac{\alpha_f}{v^i_{x_i,f}}$. Thus the total runtime for the update drops to $O(km)$.

This analysis shows that PARAFAC scales much better than TD with respect to m and k. Even though PARAFAC and TD have similar model equations, by setting the core tensor constant and diagonal the complexity drops largely because the interactions between factors are reduced to interactions within the same dimension.

5.2.4 Two-Mode PARAFAC

In the case of two modes ($m=2$), PARAFAC is identical to the popular Matrix Factorization model (eq. (5.11)):

$$\hat{y}^{\text{PARAFAC}}_{x_1,x_2} = \sum_{f=1}^{k} \prod_{i=1}^{2} v^i_{x_i,f} = \sum_{f=1}^{k} v^1_{x_1,f} \cdot v^2_{x_2,f} = \langle \mathbf{v}^1_{x_1}, \mathbf{v}^2_{x_2} \rangle = \hat{y}^{\text{MF}}_{x_1,x_2} \tag{5.20}$$

That means that PARAFAC is a higher-order counter part of matrix factorization. In the following, we will present another model which is also a higher-order counter part of MF.

5.3 Pairwise Interaction Tensor Factorization (PITF)

Instead of modelling an m-ary relation directly with one m-ary product like PARAFAC, PITF models many pairwise relations. The idea is to model one interaction explicitly for each variable pair (Rendle and Schmidt-Thieme, 2010). In general, there are $\frac{m(m-1)}{2} \in O(m^2)$ distinct pairs of variables. For higher order problems (i.e. large m), modelling all pairwise interactions independently is not feasible. But, we will show that for learning ranking problems and optimizing with BCR, both the prediction and learning is invariant to many of these interactions and thus they can be dropped. In total, this will lead to only $(m-1) \in O(m)$ pairwise interactions that are relevant for ranking. Furthermore, we will show that PITF is a special case of PARAFAC and thus also of TD.

5.3.1 Model Equation

The general model equation for PITF without pruning is:

$$\hat{y}^{\text{G-PITF}}_{x_1,\ldots,x_m} := \sum_{i=1}^{m}\sum_{j=i+1}^{m} \langle \mathbf{v}^{i,j}_{x_i}, \mathbf{v}^{j,i}_{x_j} \rangle = \sum_{i=1}^{m}\sum_{j=i+1}^{m}\sum_{f=1}^{k_{i,j}} v^{i,j}_{x_i,f} \cdot v^{j,i}_{x_j,f} \tag{5.21}$$

with model parameters:

$$V^{i,j} \in \mathbb{R}^{|X_i| \times k_{i,j}}, \; V^{j,i} \in \mathbb{R}^{|X_j| \times k_{i,j}}, \quad \forall i,j \in \{1,\ldots,m\},\, i > j \tag{5.22}$$

That means, there are $\frac{m(m-1)}{2}$ factorization pairs.

5.3.1.1 Relationship to PARAFAC and TD

Next, we will discuss the relationship between PITF and both PARAFAC and TD. Figure 5.3 shows how a general 3-mode PITF model can be expressed by the Tucker Decomposition.

Lemma 5.1 (PITF is a special case of PARAFAC). *Every PITF model can be expressed by a PARAFAC model with $k = \sum_{i=1}^{m}\sum_{j=i+1}^{m} k_{i,j}$ dimensions.*

Proof. Assume, an arbitrary PITF model is given and let $k_{i,j}$ be its pairwise factor dimensionality. The corresponding PARAFAC model of order $k = \sum_{i=1}^{m}\sum_{j=i+1}^{m} k_{i,j}$ can be generated from the PITF model by setting parts of the feature matrices to constant 1. To enhance readability of this proof, we denote parameters of the PITF model by U and the parameters of the PARAFAC model by V.

Let Z_p be the set of all ordered pairs:

$$Z_p := \{(i,j) \,|\, i,j \in \{1,\ldots,m\}, i > j\} \tag{5.23}$$

It is known that $|Z_p| = \frac{m(m-1)}{2}$. Now let Z be the set of all ordered index pairs with indices over their factorization dimensionality in PITF:

$$Z := \{(i,j,f) \,|\, (i,j) \in Z_p, f \in \{1,\ldots,k_{i,j}\}\} \tag{5.24}$$

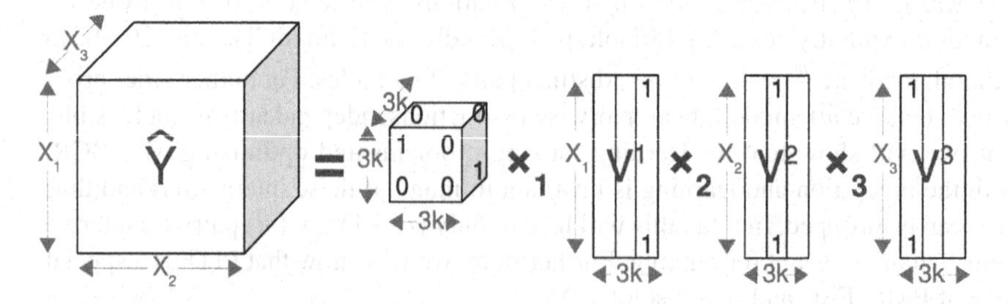

Fig. 5.3 General PITF: Like in PARAFAC, the core tensor is diagonal, but in G-PITF one third of each factor matrix V_i is fixed to constant 1. This results in three pairwise interactions: (x_1,x_2), (x_1,x_3) and (x_2,x_3).

The cardinality of Z is the size of the corresponding PARAFAC model:

$$|Z| = k = \sum_{i=1}^{m} \sum_{j=i+1}^{m} k_{i,j} \tag{5.25}$$

Let ϕ be a bijective mapping from Z to $\{1,\ldots,|Z|\}$. With this, we can rewrite the PARAFAC model:

$$\hat{y}^{\text{PARAFAC}}_{x_1,\ldots,x_m} = \sum_{f=1}^{k} \prod_{i=1}^{m} v^i_{x_i,f} = \sum_{(i,j,f)\in Z} \prod_{l=1}^{m} v^l_{x_l,\phi(i,j,f)} \tag{5.26}$$

Now, we set some of the parameters of the PARAFAC model to constants:

$$v^l_{x_l,\phi(i,j,f)} = \begin{cases} u^{i,j}_{x_i,f} & \text{if } l = i \\ u^{j,i}_{x_j,f} & \text{if } l = j \\ 1, & \text{else} \end{cases} \tag{5.27}$$

Subsituted in eq. (5.26) this leads to:

$$\begin{aligned} \hat{y}^{\text{PARAFAC}}_{x_1,\ldots,x_m} &= \sum_{(i,j,f)\in Z} \prod_{l=1}^{m} v^l_{x_l,\phi(i,j,f)} = \sum_{(i,j,f)\in Z} u^{i,j}_{x_i,f}\, u^{j,i}_{x_j,f} = \sum_{i=1}^{m} \sum_{j=i+1}^{m} \sum_{f=1}^{k_{i,j}} u^{i,j}_{x_i,f}\, u^{j,i}_{x_j,f} \\ &= \sum_{i=1}^{m} \sum_{j=i+1}^{m} \langle \mathbf{u}^{i,j}_{x_i}, \mathbf{u}^{j,i}_{x_j} \rangle = \hat{y}^{\text{G-PITF}}_{x_1,\ldots,x_m} \end{aligned} \tag{5.28}$$

Note that this proofs shows that every PITF model can be expressed by a PARAFAC model. It is important to note that the contradiction does not hold, i.e. a general PARAFAC model of mode $m \neq 2$ can not be represented by a PITF model.

5.3.1.2 PITF for Ranking

The general PITF model eq. (5.21) requires a large number of parameters, as the number of variable pairs growths quadratic in m. Next, we show that when PITF is used for ranking with respect to the variable domain X_m, many pairwise interactions can be dropped without loosing expressiveness.

The model equation for ranking PITF is:

$$\hat{y}^{\text{PITF}}_{x_1,\ldots,x_m} := \sum_{i=1}^{m-1} \langle \mathbf{v}^{i,m}_{x_i}, \mathbf{v}^{m,i}_{x_m} \rangle \tag{5.29}$$

with model parameters:

$$V^{i,m} \in \mathbb{R}^{|X_i| \times k_{i,m}},\ V^{m,i} \in \mathbb{R}^{|X_m| \times k_{i,m}},\quad \forall i \in \{1,\ldots,m-1\} \tag{5.30}$$

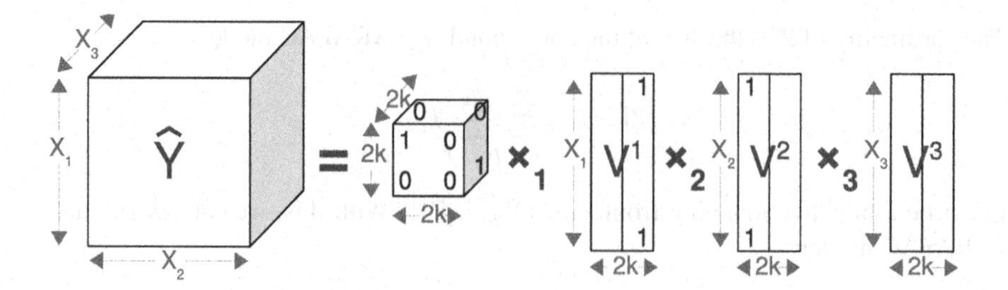

Fig. 5.4 PITF for Ranking: From the general PITF, the interaction (x_1, x_2) can be removed as it has no influence on ranking with respect to x_3. This holds only for predicting rankings and optimizing the parameters for ranking (like with BCR-OPT).

Before showing that this model is equivalent to the general PITF model for the task of ranking, we first note the general expressiveness and the relationship to PARAFAC and TD. (i) It is clear, that every ranking PITF model is also a general PITF model. This can be seen by setting $k_{i,j} = 0, \forall i \neq m, j \neq m$. And (ii) thus also every PITF is a special case of PARAFAC and TD. An example for three modes is shown in figure 5.4.

Lemma 5.2 (Invariance of Pairs). *The G-PITF (eq. (5.21)) and PITF (eq. (5.29)) model are invariant under ranking (eq. (3.34)) and learning with respect to BCR (eq. (4.11)).*

Proof. G-PITF can be rewritten as:

$$\hat{y}^{\text{G-PITF}}_{x_1,\ldots,x_m} = \hat{y}^{\text{PITF}}_{x_1,\ldots,x_m} + z_{x_1,\ldots,x_{m-1}} \tag{5.31}$$

with

$$z_{x_1,\ldots,x_{m-1}} := \sum_{i=1}^{m-1} \sum_{j=i+1}^{m-1} \langle \mathbf{v}^{i,j}_{x_i}, \mathbf{v}^{j,i}_{x_j} \rangle \tag{5.32}$$

Now G-PITF consists of a term PITF that depends on x_m and a term $z_\mathbf{c}$ that depends not on x_m but only on the context $\mathbf{c} = (x_1, \ldots, x_{m-1})$. The equivalence of PITF and G-PITF with respect to ranking and learning with BCR is proofed by the following stronger lemma.

Lemma 5.3 (Invariance of Additive Terms). $\hat{y}_{x_1,\ldots,x_m}$ *and* $\hat{y}'_{x_1,\ldots,x_m} := \hat{y}_{x_1,\ldots,x_m} + z_\mathbf{c}$ *are invariant to ranking and learning with BCR.*

Proof. First, we show the invariance to ranking. Let $\succ$ be the ranking induced by $\hat{y}$ and $\succ'$ be the ranking induced by $\hat{y}'$:

$$\begin{aligned}
x_i \succ_{\mathbf{c}} x_j &\overset{eq.(3.34)}{\Leftrightarrow} \hat{y}_{\mathbf{c},x_i} > \hat{y}_{\mathbf{c},x_j} \\
&\Leftrightarrow \hat{y}_{\mathbf{c},x_i} + z_{\mathbf{c}} > \hat{y}_{\mathbf{c},x_j} + z_{\mathbf{c}} \\
&\Leftrightarrow \hat{y'}_{\mathbf{c},x_i} > \hat{y'}_{\mathbf{c},x_j} \overset{eq.(3.34)}{\Leftrightarrow} x_i \succ'_{\mathbf{c}} x_j
\end{aligned}$$

Secondly, we show the invariance for learning with BCR. Recall the definition of BCR-OPT:

$$\begin{aligned}
&\underset{\Theta}{\operatorname{argmax}}\, \text{BCR-OPT} \\
=&\underset{\Theta}{\operatorname{argmax}} \sum_{(\mathbf{c},x_i,x_j)\in\mathscr{C}\times X_m^2} d_s(\mathbf{c},x_i,x_j) \ln\sigma(\hat{y}(\mathbf{c},x_i) - \hat{y}(\mathbf{c},x_j)) - \sum_{\theta\in\Theta} \frac{1}{2}\lambda_\theta \theta^2
\end{aligned}$$

The model $\hat{Y}$ is only used as the difference of two variable instances $x_i, x_j \in X_m$. Thus:

$$\begin{aligned}
\hat{y}'(\mathbf{c},x_i) - \hat{y}'(\mathbf{c},x_j) &= \hat{y}'_{\mathbf{c},x_i} - \hat{y}'_{\mathbf{c},x_j} = \hat{y}_{\mathbf{c},x_i} + z_{\mathbf{c}} - \hat{y}_{\mathbf{c},x_j} - z_{\mathbf{c}} \\
&= \hat{y}_{\mathbf{c},x_i} - \hat{y}_{\mathbf{c},x_j} = \hat{y}(\mathbf{c},x_i) - \hat{y}(\mathbf{c},x_j)
\end{aligned}$$

So the additional term $z_{\mathbf{c}}$ vanishes completely in the optimization. That means it is never used in the optimization function and thus it is also not updated with gradient descent learning. In total, as the term $z_{\mathbf{c}}$ is not regarded, the models $\hat{Y}$ and $\hat{Y}'$ are invariant.

Note that this only holds for optimization with a ranking criterion like BCR. For other optimization approaches like element-wise losses (e.g. LSE) the invariance does not hold and thus the interactions can not be dropped. The reason is, that these approaches try to estimate an absolute value and this value (largely) depends on the context.

5.3.2 *Gradients*

For learning a PITF model with gradient descent, the partial derivatives of eq. (5.29) for each model parameter with respect to an instance $\mathbf{x} = (x_1, \ldots, x_m)$ are:

$$\frac{\partial \hat{y}^{\text{PITF}}_{x_1,\ldots,x_m}}{\partial v^{i,j}_{x_i,f}} = v^{j,i}_{x_j,f}, \quad \frac{\partial \hat{y}^{\text{PITF}}_{x_1,\ldots,x_m}}{\partial v^{j,i}_{x_j,f}} = v^{i,j}_{x_i,f} \tag{5.33}$$

where for general PITF:

$$i, j \in \{1, \ldots, m\}, \quad i < j \tag{5.34}$$

and for ranking PITF:

$$i \in \{1, \ldots, m-1\}, \quad j = m \tag{5.35}$$

5.3.3 Complexity

Next, we discuss the complexity of the ranking PITF model in terms of number of free parameters, and number of operations for prediction and update:

- **Number of free parameters:** In ranking PITF, there is one factor matrix for each variable but for the variable to rank, there are $m-1$ independent factor matrices. In total, the number of free parameters is:

$$\sum_{i=1}^{m-1} k_{i,m} \left(|X_i| + |X_m|\right) \tag{5.36}$$

  With the assumption that all factorization dimensionalities are equal (i.e. $\forall i \in \{1,\ldots,m-1\} : k_{i,m} \stackrel{!}{=} k$), this simplifies to:

$$k \sum_{i=1}^{m} |X_i| + k\,(m-2)\,|X_m| \tag{5.37}$$

  That means like for PARAFAC, the number of free parameters for PITF is linear in k and m. But there are $k\,(m-2)\,|X_m|$ more parameters in the PITF model than in PARAFAC, when the same k is chosen.
- **Computation of Model Equation:** Predicting one entry $\mathbf{x}$ of Y with PARAFAC (eq. (5.15)) corresponds to $(m-1)$ summations over a scalar product of size k. Thus the number of operations is:

$$(m-1)\,(2k-1) \in O(km) \tag{5.38}$$

- **Computation of the Gradients:** Given a relation instance $\mathbf{x}$, all factors of the related variable instances are updated. That means, there are $km + k\,(m-2)$ parameters to update. The gradient in eq. (5.33) requires no computation time, so the number of basic operations is:

$$2k\,(m-1) \in O(km) \tag{5.39}$$

  Again, this is linear in both k and m.

5.3.4 Two-Mode PITF

Again for two-mode scenarios, general PITF and PITF for ranking is identical to matrix factorization. The reason is, that for two modes, there is only one interaction and thus:

$$\hat{y}^{\text{PITF}}_{x_1,x_2} := \langle \mathbf{v}^{1,2}_{x_1}, \mathbf{v}^{2,1}_{x_2} \rangle = \hat{y}^{\text{MF}}_{x_1,x_2} \tag{5.40}$$

Furthermore this shows that PITF and PARAFAC have the same model equation for two-mode problems.

Table 5.1 Model complexity with respect to number of free parameters and computation runtime for prediction and gradients.

Model	Free Parameters	Prediction	Gradient
TD	$k\sum_{i=1}^{m}\lvert X_i\rvert + k^m$	mk^m	mk^m
PARAFAC	$k\sum_{i=1}^{m}\lvert X_i\rvert$	mk	mk
PITF	$k\sum_{i=1}^{m}\lvert X_i\rvert + k(m-2)\lvert X_m\rvert$	mk	mk

5.4 Expressiveness

Now, we want to summarize the expressiveness of the approaches. We use $\mathcal{M}$ to denote the model classes and we write $\mathcal{M}^{\text{TD}}$ for Tucker Decomposition, $\mathcal{M}^{\text{PARAFAC}}$ for PARAFAC and $\mathcal{M}^{\text{PITF}}$ for PITF. Let $\mathcal{M}(m)$ denote the model class for a specific number of modes.

Two-Mode Problems

We have seen, that $\mathcal{M}^{\text{PARAFAC}}(2) = \mathcal{M}^{\text{PITF}}(2)$ and also that they are equivalent to usual matrix factorization $\mathcal{M}^{\text{MF}}$. Furthermore, we have shown, that $\mathcal{M}^{\text{TD}}(2) \subset \mathcal{M}^{\text{MF}}$ because the core matrix of TD allows all possible interactions between factors of different dimensions. In total, we have:

$$\mathcal{M}^{\text{TD}}(2) \supset \mathcal{M}^{\text{PARAFAC}}(2) = \mathcal{M}^{\text{PITF}}(2) = \mathcal{M}^{\text{MF}} \tag{5.41}$$

Higher-order Problems ($m \geq 3$)

The equivalence of PITF and PARAFAC only holds for $m = 2$. For higher dimensions, PITF cannot express PARAFAC as it can only model two way interactions (multiplications) directly. But we have seen, that every PITF model can be formulated as a PARAFAC model. And furthermore, TD subsumes PARAFAC for any number of modes ($m \geq 2$). To summarize, the expressiveness is:

$$\mathcal{M}^{\text{TD}}(m) \supset \mathcal{M}^{\text{PARAFAC}}(m) \supset \mathcal{M}^{\text{PITF}}(m), \quad \forall m \geq 3 \tag{5.42}$$

Discussion

At first glance, one might think that reducing the expressiveness helps only in terms of runtime. E.g. PARAFAC has a better prediction and update complexity than TD (see table 5.1). If this would be the case, PITF models would not make any sense as PARAFAC subsumes them and both have the same computational complexity. But actually, choosing a model with less expressiveness does not have to lead to worse prediction quality — i.e. that quality is traded in for e.g. runtime. We will show this in detail for tag recommenders (chapter 7). The reason is that e.g. PITF explicitly models a structure that might be hard to find under sparsity for the TD and PARAFAC approach. Especially, regularization approaches like ridge regression (see eq. (4.11)) which usually assume that the model parameters are normally distributed with mean zero $\theta \sim \mathcal{N}\left(0, \frac{1}{2\lambda_\theta}\right)$ might fail to learn the structure modeled explicitly. Thus, if a model structure is known a priori, it might be better to model it explicitly than trying to learn it. This is especially important for our problem settings with large sparsity.

5.5 Computational Aspects

Finally, the costs of factorization models are discussed from a practical point of view.

Memory usage

A huge advantage of factorization models is that they store for each variable instance (e.g. each user) only a small number of factors (k). Such a small factor vector is the only information needed to be stored (e.g. for a customer or a product) to make predictions. Other information like the historical events (S) or user/ item attributes (like name, brand) are not necessary for predictions. Clearly, storing a small number of factors for each variable instance is feasible for any real world system – e.g. online shops typically have to store already much more information per customer or product (like name, age, address). Furthermore, factorization models allow to control the trade-off between quality and memory consumption by increasing or decreasing the number of latent factors (k).

Runtime

For the linear models (PARAFAC and PITF), prediction is fast as it only depends linearly on the precomputed factors. Again, the prediction is independent of any other information like the historical events (S) or other information about the variable instances (e.g. the name of customer, the genre of a book). This is in contrast to other models like kNN where the prediction depends on the historical events.

On the other hand, the fast prediction of factorization models comes to the prize of training the model. But in an application, training is usually done offline. Furthermore, PARAFAC and PITF are easily parallelizable because their model equation of two instances $\mathbf{x}$ and $\mathbf{x}'$ does not share any parameters if $x_i \neq x'_i, \forall i \in \{1,\ldots,m\}$ – this holds for most instances when the domains are large. Additionally, the models can also be trained with online-updates. This is important in practice because usually a model is trained offline, then it is deployed and updates should be performed online. We have provided such a online-update algorithm for regularized least-square matrix factorization (Rendle and Schmidt-Thieme, 2008) – a similar algorithm can be used for BCR learning of PARAFAC or PITF. In the evaluation of (Rendle and Schmidt-Thieme, 2008), online updating the factors of a user or item on the large Netflix dataset ($|S| \approx 100,000,000$) costs between 0ms and 15ms (depending on the number of historical events for this user/ item). Moreover, it was shown that for variable instances with a large history, online-updating after each event is not necessary.

In all this shows that factorization models are highly applicable in practice because their memory storage is small, the predictions are fast and learning can be done by online-updating the factors in real-time.

References

Harshman, R.A.: Foundations of the parafac procedure: models and conditions for an 'exploratory' multimodal factor analysis. UCLA Working Papers in Phonetics, 1–84 (1970)

Rendle, S., Schmidt-Thieme, L.: Online-updating regularized kernel matrix factorization models for large-scale recommender systems. In: RecSys 2008: Proceedings of the 2008 ACM Conference on Recommender Systems, pp. 251–258. ACM, New York (2008)

Rendle, S., Schmidt-Thieme, L.: Pairwise interaction tensor factorization for personalized tag recommendation. In: WSDM 2010: Proceedings of the third ACM International Conference on Web search and Data Mining, pp. 81–90. ACM, New York (2010)

Srebro, N., Rennie, J.D.M., Jaakola, T.S.: Maximum-margin matrix factorization. In: Advances in Neural Information Processing Systems, vol. 17, pp. 1329–1336. MIT Press, Cambridge (2005)

Tucker, L.: Some mathematical notes on three-mode factor analysis. Psychometrika 31, 279–311 (1966)

Part III
Application

In this part, we apply the theory that has been developed in part II. We investigate three scenarios for context-aware ranking.

The first scenario is personalization, where items should be ranked individually for each user. The feedback of the user is implicit, e.g. the feedback corresponds to the users actions in the past. This is a very important recommendation scenario with applications to e.g. online shopping (customers buying products), DVD rental (users renting DVDs) or IP-TV (users watching movies). Here, we will compare our context-aware ranking methods to other state-of-the-art approaches like weighted regularized matrix factorization or k-nearest neighbour.

Secondly, the theory of context-aware ranking is applied to tag recommendation. In this case, the context consists of users and items (e.g. bookmarks, songs) and tags should be recommended. Tagging plays an important role in the Web 2.0 and tag recommenders help the user in the annotation process. Here, we apply the tensor factorization models of chapter 5 and compare them to the competing approaches Folkrank, Pagerank and HOSVD.

Finally, we investigate a scenario with time information. The data consists of sequences of sets. For example, the shopping carts of a user over time can be seen as a sequence of sets, where each shopping cart is one set. The context in this case is the user and the time. And thus the ranking of the items (e.g. products) should be both user and time specific. For capturing the time interactions, we develop a personalized Markov chain model where the transitions are factorized to improve the parameter estimation. We show that the standard matrix factorization model is a special case of our factorized personalized Markov chain.

Chapter 6
Item Recommendation

Recommending content is an important task in many information systems. For example online shopping websites like Amazon give each customer personalized recommendations of products that the user might be interested in. Other examples are video portals like YouTube that recommend videos to visitors. Personalization is attractive both for content providers, who can increase sales or views, and for customers, who can find interesting content more easily. In this chapter, we focus on item recommendation where the task is to create a user-specific ranking for a set of items. Preferences of users about items are learned from the user's past interaction with the system – e.g. his buying history, viewing history, etc. Thus, the context in item recommenders is the user and user-aware rankings should be generated.

Recommender systems are an active topic of research. Most recent work is on scenarios where users provide explicit feedback, e.g. in terms of ratings. Nevertheless, in real-world scenarios most feedback is not explicit but implicit. Implicit feedback is tracked automatically, like monitoring clicks, view times, purchases, etc. Thus it is much easier to collect, because the user has not to express his taste explicitly. In fact implicit feedback is already available in almost any information system – e.g. web servers record any page access in log files.

We start with introducing the related work in the field of item recommendation. Then, we discuss the problem setting and show how it fits into the context-aware ranking framework. Afterwards, we apply the *Bayesian Context-aware Ranking* (BCR) framework to personalization, thus we get *Bayesian Personalized Ranking* (BPR). We show the optimization criterion and learning algorithm for this instance of BCR. Then, we describe how the two popular item recommendation models k-nearest neighbor and matrix factorization can be learned with BPR. In the evaluation, we compare the quality of these approaches to the state-of-the-art recommenders of weighted regularized matrix factorization and cosine kNN.

6.1 Related Work

The most popular model for recommender systems is k-nearest neighbor (kNN) collaborative filtering (Deshpande and Karypis, 2004). Traditionally, the similarity

S. Rendle: Context-Aware Ranking with Factorization Models, SCI 330, pp. 69–84.
springerlink.com

matrix of kNN is computed by heuristics – e.g. the Pearson correlation – but in recent work (Koren, 2008) the similarity matrix is treated as model parameters and is learned specifically for the task. Recently, matrix factorization (MF) has become very popular in recommender systems both for implicit and explicit feedback. In early work (Sarwar et al, 2002) singular value decomposition (SVD) has been proposed to learn the feature matrices. MF models learned by SVD have shown to be very prone to overfitting (Kurucz et al, 2007). Thus regularized learning methods have been proposed. For item prediction Hu et al (2008) and Pan et al (2008) propose a regularized least-square optimization with case weights (WR-MF). The case weights can be used to reduce the impact of negative examples. Hofmann (2004) proposes a probabilistic latent semantic model for item recommendation. Schmidt-Thieme (2005) converts the problem into a multi-class problem and solves it with a set of binary classifiers.

Even though all the work on item prediction discussed above is evaluated on personalized ranking datasets, none of these methods directly optimizes its model parameters for ranking. Instead they optimize to predict if an item is selected by a user or not. In our work we derive an optimization criterion for personalized ranking that is based on pairs of items (i.e. the user-specific order of two items). We will show how state-of-the-art models like MF or adaptive kNN can be optimized with respect to this criterion to provide better ranking quality than with usual learning methods. A detailed discussion of the relationship between our approach and the WR-MF approach of Hu et al (2008) and Pan et al (2008) as well as maximum margin matrix factorization (Srebro et al, 2005; Weimer et al, 2008) can be found in section 6.5.

In this paper, we focus on offline learning of the model parameters. Extending the learning method to online learning scenarios – e.g. a new user is added and his history increases from 0 to 1, 2, ... feedback events – has already been studied for MF for the related task of rating prediction Rendle and Schmidt-Thieme (2008). The same fold-in strategy can be used for BPR.

There is also related work on learning to rank with non-collaborative models. One direction is to model distributions on permutations (Kondor et al, 2007; Huang et al, 2008). Burges et al (2005) optimize a neural network model for ranking using gradient descent. All these approaches learn only one ranking – i.e. they are non-personalized. In contrast to this, our models are collaborative models that learn personalized rankings, i.e. one individual ranking per user. In our evaluation, we show empirically that in typical recommender settings our personalized BPR model outperforms even the theoretical upper bound for non-personalized ranking.

6.2 Personalized Ranking from Implicit Feedback

The task of personalized ranking is to provide a user with a ranked list of items. This is also called item recommendation. An example is an online shop that wants to recommend a personalized ranked list of items that the user might want to buy.

In this work, we investigate scenarios where the ranking has to be inferred from the implicit behavior (e.g. purchases in the past) of the user. Interesting about implicit feedback systems is that only positive observations are available. The non-observed user-item pairs – e.g. a user has not bought an item yet – are a mixture of real negative feedback (the user is not interested in buying the item) and missing values (the user might want to buy the item in the future).

6.2.1 Formalization

For easier readability, we adapt the formalization of chapter 3 to the special case of item recommendation, e.g. we use the more meaningful symbols U for users and I for items instead of X_1 and X_2.

Item recommendation is a two mode problem ($m = 2$). Let $U = X_1$ be the set of all users and $I = X_2$ the set of all items. In our scenario, the observed training data is binary set data – i.e. it can be written as $S \subseteq U \times I$ (see the topmost matrix in figure 6.1). Examples for such feedback are purchases in an online shop, views in a video portal or clicks on a website. The task of the recommender system is now to provide the user with a personalized ranking $\succ_u \subset I^2$ of all items.

6.2.2 Analysis of the Problem Setting

As we have indicated before, in implicit feedback systems only positive classes are observed. The remaining data is a mixture of actually negative and missing values. The most common approach for dealing with the missing value problem is to ignore all of them but then typical machine learning models are unable to learn anything, because they cannot distinguish between the two levels anymore. Machine learning approaches for item recommenders (Hu et al, 2008; Pan et al, 2008) typically create the training data from S by giving all pairs $(u,i) \in S$ a positive class label and all other combinations in $(U \times I) \setminus S$ a negative one – that means they try to reconstruct the observed data S of the past. As we have discussed in chapter 3, the problem with this approach is that all elements the model should rank in the future ($(U \times I) \setminus S$) are presented to the learning algorithm as negative feedback during training. That means a model with enough expressiveness (that can fit the training data exactly) cannot rank at all as it predicts only negative class values. The only reason why such machine learning methods can predict rankings are strategies to prevent overfitting, like regularization.

Instead, our approach is to see the task as a ranking problem rather than a classification problem and to create pairwise training data D_S (see chapter 3). This process is illustrated in figure 6.1. If an item i has been viewed by user u – i.e. $(u,i) \in S$ – then we assume that the user prefers this item over all other non-observed items. E.g. in Figure 6.1 user u_1 has viewed item i_2 but not item i_1, so we assume that this user prefers item i_2 over i_1: $i_2 \succ_{u_1} i_1$. For items that have both been seen by a user,

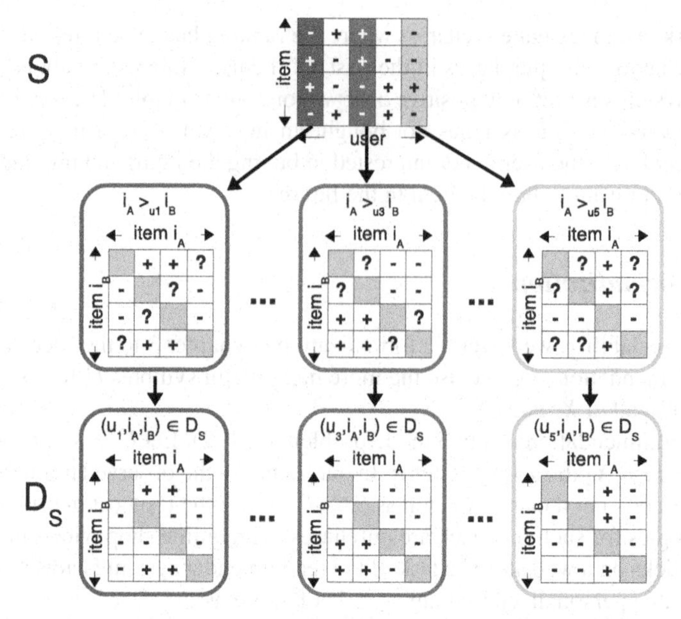

Fig. 6.1 Item Recommendation: the observed data S is binary (*top*). For each user, we try to find indication in the data S if the user prefers one item over the other (*middle*). The inferred training data D_S is then shown at the bottom. In all figures, a '+' means that the element is in the set, a '-' means it is not in the set and a '?' means that it is unknown whether or not it is in the set.

we cannot infer any preference. The same is true for two items that a user has not seen yet (e.g. item i_1 and i_4 for user u_1). Now, D_S can be created according to eq. (3.21), which reads the following for item recommendation:

$$D_S := \{(u,i,j) : (u,i) \in S \wedge (u,j) \notin S\} \tag{6.1}$$

Note that even though D_S explicitly contains only positive pairs, it also contains implicitly the negative ones because pairs are antisymmetric – but it does not contain the pairs that are neither positive nor negative (missing values), see figure 6.1.

In total, our approach has two advantages:

1. The pairwise training data consists of both positive and negative pairs and missing values. The missing values between two non-observed items are exactly the item pairs that have to be ranked in the future. That means from a pairwise point of view the training data D_S and the test data is disjoint. This does not hold for the classification approach that tries to reconstruct S and $(U \times I) \setminus S$.
2. The training data is created for the actual objective of ranking, i.e. the observed subset D_S of $\succ_u$ is used as training data.

6.3 Learning Personalized Ranking

In this section, we transfer the general BCR method to item recommendation. As we have stated before, for item recommendation, the context is the user, i.e. $\mathscr{C} = U$. Thus, we refer to Bayesian Context-aware Ranking (BCR) as *Bayesian Personalized Ranking* (BPR). We will both present the optimization criterion BPR-OPT and the learning algorithm BPR-LEARN.

6.3.1 *Optimization Criterion (*BPR-OPT*)*

Analogous to BCR (see section 4.1), we can derive BPR-OPT which is the MAP estimator for the model parameters Θ of an estimator $\hat{y}$:

$$\operatorname*{argmax}_{\Theta} p(\Theta \mid \succ) = \operatorname*{argmax}_{\Theta} p(\succ \mid \Theta)\, p(\Theta) \tag{6.2}$$

where the probability for $\succ$ is:

$$p(\succ \mid \Theta) = \prod_{(u,i,j) \in U \times I^2} p(i \succ_u j \mid \Theta)^{2\,\delta(i \succ_u j)} \tag{6.3}$$

Again, we assume that $\hat{y}$ is used as an estimator for $\succ$ (see section 4.1.3) and thus:

$$p(i \succ_u j | \Theta) := \sigma(\hat{y}(u,i) - \hat{y}(u,j)) \tag{6.4}$$

And for the prior, we assume a normal distribution, thus:

$$p(\Theta) = \prod_{\theta \in \Theta} \sqrt{\frac{\lambda_\theta}{\pi}} \exp\left(-\lambda_\theta\, \theta^2\right) \tag{6.5}$$

Putting everything together, leads to BPR-OPT which is defined as the MAP estimator given the training data:

$$\operatorname*{argmax}_{\Theta} \text{BPR-OPT} := \operatorname*{argmax}_{\Theta} \sum_{(u,i,j) \in D_S} \ln \sigma(\hat{y}(u,i) - \hat{y}(u,j)) - \sum_{\theta \in \Theta} \frac{1}{2} \lambda_\theta\, \theta^2 \tag{6.6}$$

where λ_θ is a model specific regularization parameter.

6.3.2 *Learning Algorithm (*BPR-LEARN*)*

Next, we specialize BCR-LEARN to item recommendation to derive BPR-LEARN. This algorithm is based on stochastic gradient descent where cases $(u,i,j) \in D_S$ are drawn by bootstrapping. The benefits of a stochastic approach with bootstrapping over the standard approach of computing the full gradient or user-wise stochastic descent are discussed in section 4.2.1.

Gradients

For performing gradient descent, the partial derivative of BPR-OPT with respect to the model parameters θ given a case $(u,i,j) \in D_S$ is:

$$\frac{\partial}{\partial \theta}\text{BPR-OPT} = \delta_{u,i,j} \frac{\partial}{\partial \theta}(\hat{y}(u,i) - \hat{y}(u,j)) - \lambda_\theta \, \theta \tag{6.7}$$

with:

$$\delta_{u,i,j} := (1 - \sigma(\hat{y}(u,i) - \hat{y}(u,j))) \tag{6.8}$$

Drawing of Cases

It is not feasible to explicitly enumerate all triples in D_S. Instead, we can draw cases (u,i,j) from D_S without enumerating them. This can be done by first drawing a tuple (u,i) from S and then drawing a negative item j. The drawing of a negative item can be performed by (1) drawing j from I and (2) rejecting it if $(u,j) \in S$. Rejection is very unlikely because most items are not purchased/ bought/ etc. by the user. More details about this can be found in section 4.2.3.

BPR-Learn

Algorithm 2 shows the learning method BPR-LEARN for optimizing the model parameters Θ with respect to BPR-OPT. This algorithm is an adaption of BCR-LEARN (see algorithm 1) to the context of personalization. After initializing the model parameters Θ, the parameters are learned iteratively. First a case is drawn as described before and then gradient descent is performed on the related parameters. For termination of the main-loop, we searched the best number of iterations on a holdout set. This corresponds to early stopping and can be seen as an additional way of regularizing.

6.4 Item Recommendation Models

In the following, we describe two state-of-the-art model classes for item recommendation and show how they can be learned with our proposed BPR methods. We have chosen the two diverse model classes of matrix factorization (Hu et al, 2008; Rennie and Srebro, 2005) and learned k-nearest-neighbor (Koren, 2008). Both classes try to model the hidden preferences of a user on an item. Their prediction is a real number $\hat{y}(u,i)$ per user-item-pair (u,i). As both U and I are categorical domains, one can express $y : U \times I \to \mathbb{R}$ as a matrix $Y \in \mathbb{R}^{|U| \times |I|}$. Thus, we will use this notation in the following and write $y_{u,i} := y(u,i)$.

It is important to note that even though we use the same models as in other work, we optimize them against another criterion. This will lead to a better ranking

Algorithm 2 BPR-LEARN

Input: training data S, learning rate α, regularization parameters λ_θ
Output: model parameters Θ
1: initialize Θ from $\mathcal{N}(0,\sigma^2)$
2: **repeat**
3: draw (u,i) uniformly from S
4: draw j from $\{l : (u,l) \notin S\}$
5: $\delta_{u,i,j} \leftarrow (1-\sigma(\hat{y}(u,i)-\hat{y}(u,j)))$
6: **for** $\theta \in \Theta$ **do**
7: $\theta \leftarrow \theta + \alpha\left(\delta_{u,i,j}\frac{\partial}{\partial\theta}(\hat{y}(u,i)-\hat{y}(u,j)) - \lambda_\theta\,\theta\right)$
8: **end for**
9: **until** convergence
10: **return** Θ

because our criterion is optimal for the ranking task. It does not try to regress a single predictor $\hat{y}_{u,i}$ to a single number but instead tries to classify the ranking of two predictions, i.e. $\hat{y}_{u,i} > \hat{y}_{u,j}$.

6.4.1 Matrix Factorization

In chapter 5, we have presented three tensor factorization models and have shown that for two mode tensors, PARAFAC and PITF corresponds to matrix factorization. In this section, we show how the matrix factorization model can be used for item recommendation.

The problem of predicting $\hat{y}_{u,i}$ can be seen as the task of estimating a real valued matrix $Y \in \mathbb{R}^{|U|\times|I|}$. With matrix factorization the target matrix Y is approximated by the matrix product of two low-rank matrices $W \in \mathbb{R}^{|U|\times k}$ and $H \in \mathbb{R}^{|I|\times k}$:

$$\hat{Y} := W \cdot H^t$$

where k is the dimensionality/rank of the approximation. Each vector/ row $\mathbf{w}_u$ in W can be seen as a feature vector describing a user u and similarly each row $\mathbf{h}_i$ of H describes an item i. Thus the prediction formula can also be written as:

$$\hat{y}_{u,i} = \langle \mathbf{w}_u, \mathbf{h}_i \rangle = \sum_{f=1}^{k} w_{u,f} \cdot h_{i,f}$$

Besides the dot product $\langle\cdot,\cdot\rangle$ in general any kernel can be used like in (Rendle and Schmidt-Thieme, 2008). The model parameters for matrix factorization are $\Theta = (W,H)$.

In general the best approximation of $\hat{Y}$ to Y with respect to dense element-wise least-square is achieved by the singular value decomposition (SVD). For machine learning tasks, it is known that SVD overfits (Kurucz et al, 2007) and therefore many other matrix factorization methods have been proposed, including regularized least square optimization, non-negative factorization, maximum margin factorization, etc.

For the task of ranking, i.e. estimating whether a user prefers one item over another, a better approach is to optimize against the BPR-OPT criterion. This can be achieved by using our proposed algorithm BPR-LEARN. As stated before for optimizing with BPR-LEARN, only the gradient of $\hat{y}_{u,i}$ with respect to every model parameter θ has to be known. For the matrix factorization model the derivatives given a case $(u,i,j) \in D_S$ are:

$$\frac{\partial}{\partial \theta}(\hat{y}_{u,i} - \hat{y}_{u,j}) = \begin{cases} (h_{i,f} - h_{j,f}) & \text{if } \theta \text{ is } w_{u,f}, \\ w_{u,f} & \text{if } \theta \text{ is } h_{i,f}, \\ -w_{u,f} & \text{if } \theta \text{ is } h_{j,f}, \\ 0 & \text{else} \end{cases} \tag{6.9}$$

Furthermore, we use three regularization constants: one λ_W for the user features W; for the item features H we have two regularization constants, λ_{H^+} that is used for positive updates on $h_{i,f}$, and λ_{H^-} for negative updates on $h_{j,f}$.

6.4.2 *Adaptive k-Nearest-Neighbor*

In addition to the factorization models on which we are focused in this book, we describe here another modelling approach that can be applied for two-mode problems.

Nearest-neighbor methods are very popular in collaborative filtering. They rely on a similarity measure between either items (item-based) or users (user-based). In the following we describe item-based methods as they usually provide better results, but user-based methods work analogously. The idea is that the prediction for a user u and an item i depends on the similarity of i to all other items the user has seen in the past – i.e. I_u:

$$I_u := \{i : (u,i) \in S\} \tag{6.10}$$

Often only the k most similar items of I_u are regarded – the *k-nearest neighbors*. If the similarities between items are chosen carefully, one can also compare to all items in I_u. For item prediction the model of item-based k-nearest-neighbor is:

$$\hat{y}_{u,i} = \sum_{l \in I_u \wedge l \neq i} c_{i,l} \tag{6.11}$$

where $C : I \times I$ is the symmetric item-correlation/ item-similarity matrix. Hence the model parameters of kNN are $\Theta = C$.

The common approach for choosing C is by applying a heuristic similarity measure, e.g. cosine vector similarity:

$$c_{i,j}^{\text{cosine}} := \frac{|U_i \cap U_j|}{\sqrt{|U_i| \cdot |U_j|}} \tag{6.12}$$

with:

$$U_l := \{u : (u,l) \in S\} \tag{6.13}$$

A better strategy is to adapt the similarity measure C to the problem by learning it. This can be either done by using C directly as model parameters or if the number of items is too large, one can learn a factorization HH^t of C with $H \in \mathbb{R}^{|I| \times k}$. In the following and also in our evaluation we use the first approach of learning C directly without factorizing it.

Again for optimizing the kNN model for ranking, we apply the BPR optimization criterion and use the BPR-LEARN algorithm. For applying the algorithm, the gradient of $\hat{y}_{u,i} - \hat{y}_{u,j}$ with respect to the model parameters C is:

$$\frac{\partial}{\partial \theta}(\hat{y}_{u,i} - \hat{y}_{u,j}) = \begin{cases} +1 & \text{if } \theta \in \{c_{i,l}, c_{l,i}\} \wedge l \in I_u \wedge l \neq i, \\ -1 & \text{if } \theta \in \{c_{j,l}, c_{l,j}\} \wedge l \in I_u \wedge l \neq j, \\ 0 & \text{else} \end{cases}$$

We have two regularization constants, λ_+ for updates on $c_{i,l}$, and λ_- for updates on $c_{j,l}$.

6.5 Relations to Other Methods

We discuss the relations of our proposed methods for ranking to two further item recommendation models. The first one is an element-wise weighted least-square error (eq. (4.25)) and the second one uses pairwise hinge loss (eq. (4.17)).

6.5.1 Weighted Regularized Matrix Factorization (WR-MF)

Both Pan et al (2008) and Hu et al (2008) have presented a matrix factorization method for item prediction from implicit feedback. Thus the model class is the same as we have described in section 6.4.1, i.e. $\hat{Y} := WH^t$ with the matrices $W \in \mathbb{R}^{|U| \times k}$ and $H \in \mathbb{R}^{|U| \times k}$. The optimization criterion and learning method differ substantially from our approach. Their method is an adaption of an SVD, which minimizes the square-loss. Their extensions are regularization to prevent overfitting and weights in the error function to increase the impact of positive feedback. In total their optimization criterion is:

$$\sum_{u \in U} \sum_{i \in I} c_{u,i} (\langle \mathbf{w}_u, \mathbf{h}_i \rangle - \delta((u,i) \in S))^2 + \lambda ||W||_f^2 + \lambda ||H||_f^2 \tag{6.14}$$

where $c_{u,i}$ are not model parameters but apriori given weights for each tuple (u,i). Hu et al. have additional data to estimate $c_{u,i}$ for positive feedback and they set $c_{u,i} = 1$ for the rest. Pan et al. suggest to set $c_{u,i} = 1$ for positive feedback and choose lower constants for the rest.

First of all, it is obvious that this optimization is on instance level (one item) instead of pair level (two items) as BPR. Apart from this, their optimization is a least-square which is known to correspond to the MLE for normally distributed random variables. However, the task of item recommendation is actually not a regression (quantitative), but a classification (qualitative) one, so the logistic optimization is more appropriate.

A strong point of WR-MF is that it can be learned in $O(\text{iter}\,(|S|\,k^2 + k^3\,(|I| + |U|)))$ provided that $c_{u,i}$ is constant for non-positive pairs. Our evaluation indicates that BPR-LEARN usually converges after a subsample of $|D_S|$ single update steps even though there are much more triples to learn from.

6.5.2 *Maximum Margin Matrix Factorization for Ordinal Ranking*

Weimer et al (2008) use the maximum margin matrix factorization method (MMMF) (Srebro et al, 2005) for ordinal ranking. Their MMMF is designed for scenarios with explicit feedback in terms of ratings. Even though their ranking MMMF is not intended for implicit feedback datasets, one could apply it in our scenario by giving all non-observed items the 'rating' 0 and the observed ones a 1. With these modifications their optimization criterion to be minimized would be quite similar to BPR applied for matrix factorization:

$$\sum_{(u,i,j)\in D_S} \max(0, 1 - \langle \mathbf{w}_u, \mathbf{h}_i - \mathbf{h}_j \rangle) + \lambda_w ||W||_f^2 + \lambda_h ||H||_f^2 \qquad (6.15)$$

One difference is that the error functions differ – our hinge loss is smooth and motivated by the MLE. Additionally, our BPR-OPT criterion and BPR-LEARN algorithm is generic and can be applied to several models, whereas their method is specific for MF.

Besides this, their learning method for MMMF differs from our generic approach BPR-LEARN. Their learning method is designed to work with sparse explicit data, i.e. they assume that there are many missing values and thus they assume to have much less pairs than in an implicit setting. But when their learning method is applied to implicit feedback datasets, the data has to be densified like described above and the number of training pairs D_S is in $O(|S|\,|I|)$ (see section 3.3.2). Our method BPR-LEARN can handle this situation by bootstrapping from D_S.

6.6 Evaluation

In our evaluation, we compare learning with BPR to other learning approaches. We have chosen the two popular model classes of matrix factorization (MF) and k-nearest-neighbor (kNN). Regularized MF models are known to outperform (Rennie and Srebro, 2005) many other models including the Bayesian models URP (Marlin, 2004) and PLSA (Hofmann, 2004) for the related task of collaborative rating prediction. In our evaluation, the matrix factorization models are learned by three different methods, i.e. SVD-MF, WR-MF (Hu et al, 2008; Pan et al, 2008) and our BPR-MF. For kNN (see eq. (6.11)), we compare cosine vector similarity

(Cosine-kNN) to a model that has been optimized using our BPR method (BPR-kNN)[1]. Additionally, we report results for the baseline most-popular, that weights each item user-independently, e.g.: $\hat{y}_{u,i}^{\text{most-pop}} := |U_i|$. Furthermore, we give the theoretical upper bound on the AUC for any non-personalized ranking method (np_{max}).

6.6.1 Datasets

We use two datasets of two different applications. The *Rossmann* dataset is from an online drug store[2]. It contains the buying history of 10,000 users on 4000 items. In total 426,612 purchases are recorded. The task is to predict a personalized list of the items the user wants to buy next. The second dataset is the DVD rental dataset of *Netflix*[3]. This dataset contains the rating behavior of users, where a user provides explicit ratings 1 to 5 stars for some movies. As we want to solve an implicit feedback task, we removed the rating scores from the dataset. Now the task is to predict which movies a user rates. Again we are interested in a personalized ranked list starting with the movie that is most likely to be rated. For Netflix we have created a subsample of 10,000 users, 5000 items containing 565,738 rating actions. We draw the subsample such that every user has at least 10 items ($\forall u \in U : |I_u| \geq 10$) and each item has at least 10 users ($\forall i \in I : |U_i| \geq 10$).

6.6.2 Evaluation Methodology

We use the leave-one-out evaluation scheme, where we remove for each user randomly one action (one user-item pair) from his history, i.e. we remove one entry from I_u per user u. This results in a disjoint train set S_{train} and test set S_{test}. The models are then learned on S_{train} and their predicted personalized ranking is evaluated on the test set S_{test} by the average AUC statistic (see section 3.5). A higher value of the AUC indicates a better quality. The trivial AUC of a random guess method is 0.5 and the best achievable quality is 1.

We repeated all experiments 10 times by drawing new train/test splits in each round. The hyperparameters for all methods are optimized via grid search on the train/test split of the first round and afterwards the hyperparameters are kept constant in the remaining 9 repetitions.

On Netflix, the training runtime of the largest factorization models (k=128) was 174 minutes for BPR-MF (134 on Rossmann) and 117 minutes for WR-MF (97 on Rossmann). For BPR-kNN, the runtime was 68 minutes on Netflix and 139 on Rossmann.

[1] One might also apply the weighted regularized least square scheme of WR-MF to kNN and obtain 'WR-kNN'. To the best of our knowledge, this has not been done yet and there is no proposal for a feasible learning algorithm for WR-kNN like there exists for WR-MF. Without such a speedup, the WR-kNN iterates over the full matrix with $|U||I|$ entries.

[2] `http://www.rossmannversand.de/`

[3] `http://www.netflix.com/`

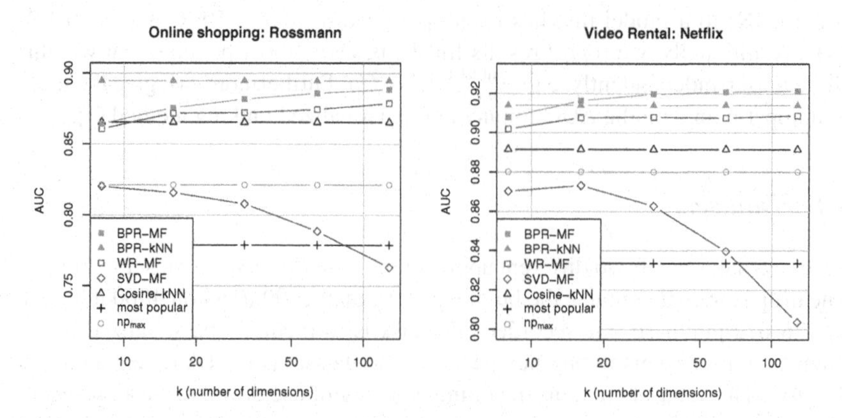

Fig. 6.2 Area under the ROC curve (AUC) prediction quality for the Rossmann dataset and a Netflix subsample. Our BPR optimization for matrix factorization BPR-MF and k-nearest neighbor BPR-kNN are compared against weighted regularized matrix factorization (WR-MF) (Hu et al, 2008; Pan et al, 2008), singular value decomposition (SVD-MF), k-nearest neighbor (Cosine-kNN) (Deshpande and Karypis, 2004) and the most-popular model. For the factorization methods BPR-MF, WR-MF and SVD-MF, the model dimensions are increased from 8 to 128 dimensions. Finally, np_{max} is the theoretical upper bound for any non-personalized ranking method.

6.6.3 Results and Discussion

Figure 6.2 shows the AUC quality of all models on the two datasets. First of all, you can see that the two BPR optimized methods outperform all other methods in prediction quality. Comparing the same models among each other one can see the importance of the optimization method. For example all MF methods (SVD-MF, WR-MF and BPR-MF) share exactly the same model, but their prediction quality differs a lot. Even though SVD-MF is known to yield the best fit on the training data with respect to element-wise least square, it is a poor prediction method for machine learning tasks as it overfits the training data. This can be seen as the quality of SVD-MF decreases with an increasing number of dimensions (expressiveness). WR-MF is a more successful learning method for the task of ranking. Due to regularization its performance does not drop but steadily rises with an increasing number of dimensions. But BPR-MF outperforms WR-MF clearly for the task of ranking on both datasets. For example on Netflix a MF model with 8 dimensions optimized by BPR-MF achieves comparable quality as a MF model with 128 dimensions optimized by WR-MF.

To summarize, our results show the importance of optimizing model parameters to the right criterion. The empirical results indicate that our BPR-OPT criterion learned by BPR-LEARN outperforms the other state-of-the-art methods for personalized ranking from implicit feedback. The results are justified by the analysis of the problem (see section 6.2.2) and by the theoretical derivation of BPR-OPT from the MLE over pairs.

6.6.4 Non-personalized Ranking

Next, we compare the AUC quality of our personalized ranking methods to the best possible non-personalized ranking method. In contrast to our personalized ranking methods, a non-personalized ranking method creates the same ranking $\succ$ for all users. We compute the theoretical upper-bound np_{max} for any non-personalized ranking method by optimizing the ranking $\succ$ on the test set S_{test}. Figure 6.2 shows that even simple personalized methods like Cosine-kNN outperform the upper-bound np_{max} — and thus also all non-personalized methods — largely.

Computation of the AUC of np_{max}

Computing a global (non-personalized) ranking $\hat{y}$ that is AUC-optimal on test is not trivial because within a user the train examples are not allowed to be rerecommended and different users have different training sizes. This leads to a slightly different weighting in the AUC computation of pair comparisons for each user. So instead of searching an optimal $\hat{y}^*$, we computed an upper-bound but non-tight estimate on the AUC score of np_{max}.

First, we define the evaluation sets of items for each user and the set of all evaluated users:

$$I_u^+ := \{i \in I : (u,i) \in S_{\text{test}} \wedge (u,i) \notin S_{\text{train}}\} \tag{6.16}$$

$$I_u^- := \{i \in I : (u,i) \notin S_{\text{test}} \wedge (u,i) \notin S_{\text{train}}\} \tag{6.17}$$

$$U^+ := \{u \in U : I_u^+ \neq \emptyset\} \tag{6.18}$$

With this, the AUC quality measure can be written as:

$$\text{AUC}(\hat{y}) = \frac{1}{|U^+|} \sum_{u \in U^+} \frac{1}{|I_u^+||I_u^-|} \sum_{i \in I_u^+} \sum_{j \in I_u^-} \delta(\hat{y}(u,i) > \hat{y}(u,j)) \tag{6.19}$$

The best non-personalized ranking on test $\hat{y}^*$ depends not on the user and has the following AUC:

$$\text{AUC}(\hat{y}^*) = \frac{1}{|U^+|} \sum_{u \in U^+} \frac{1}{|I_u^+||I_u^-|} \sum_{i \in I_u^+} \sum_{j \in I_u^-} \delta(\hat{y}^*(i) > \hat{y}^*(j)) \tag{6.20}$$

Finding the best order is non-trivial. Instead, we create an upper-bound by evaluating the contribution of a decision about a pair $(i,j) \in I^2$ on the AUC. For each pair (i,j) either i is ranked in front of j or vice versa. If the estimated ranking is $i \succ j$ then the contribution on the AUC is:

$$a_{i,j} = \frac{1}{|U^+|} \sum_{u \in U^+} \frac{1}{|I_u^+||I_u^-|} \delta(i \in I_u^+ \wedge j \in I_u^-) \tag{6.21}$$

The inverse ranking would have the contribution:

$$a_{j,i} = \frac{1}{|U^+|} \sum_{u \in U^+} \frac{1}{|I_u^+||I_u^-|} \delta(j \in I_u^+ \wedge i \in I_u^-) \tag{6.22}$$

As the ranking np_{max} has to decide if i is preferred over j, only one of the two scores can be used. This means, an upper bound estimate on the AUC is to use for each comparison the larger value:

$$AUC^{\text{upper bound}} = \frac{1}{2} \sum_{i \in I} \sum_{j \in I, i \neq j} \max(a_{i,j}, a_{j,i}) \tag{6.23}$$

Note that this is a non-tight upper bound on the AUC as the pairwise decisions might not be total (e.g. not transitive) and so this does not have to lead to a total order (ranking).

In our experiments, this non-tight upper bound is close to the true AUC value of np_{max}. This can be seen by calculating the interval where np_{max} is guaranteed to lie in. $AUC^{\text{upper bound}}$ gives an upper bound; a lower bound can be computed by ranking the items by how often they appear in the test set ('most-popular on test'). In our experiments both AUC scores are quite similar, e.g. on Netflix with most-popular on test 0.8794 vs. $AUC^{\text{upper bound}} = 0.8801$.

6.6.5 *Practical Impact*

Finally, we discuss the practical impact of the improvements. Linden et al (2003) found that in the Amazon online shop, the conversion rates[4] generated by a cosine-knn recommender *vastly exceed* the ones of the most-popular recommender (*top-seller lists*). Relating this finding with the ranking accuracy in our lab experiments (e.g. on the Netflix dataset), it would mean that the difference of cosine-knn (11% error) to most-popular (17% error) is likely to have a huge practical impact, so we might conclude that a further decrease from 11% error (cosine-knn) to 8% error (BPR-MF) has also a significant impact.

6.7 Conclusion

In this chapter, we have presented how to apply the context-aware ranking method to item recommendation. We have shown that item recommendation is a context-aware ranking problem from incomplete data. This allows to transfer the BCR-OPT and BCR-LEARN methods to personalized ranking (BPR). We have demonstrated how this generic method can be applied to the two state-of-the-art item recommender models of matrix factorization and adaptive kNN. In our evaluation, we have shown

[4] The convertion rate is one of the most important measures to assess the practical success of a recommender system.

empirically that for the task of personalized ranking, models learned by BPR outperform the same models that are optimized with respect to other criteria. Our results show that the prediction quality does not only depend on the model but also largely on the optimization criterion. Both our theoretical and empirical results indicate that the BPR optimization method is the right choice for the important task of item recommendation.

References

Burges, C., Shaked, T., Renshaw, E., Lazier, A., Deeds, M., Hamilton, N., Hullender, G.: Learning to rank using gradient descent. In: ICML 2005: Proceedings of the 22nd International Conference on Machine Learning, pp. 89–96. ACM Press, New York (2005)

Deshpande, M., Karypis, G.: Item-based top-n recommendation algorithms. In: ACM Transactions on Information Systems, vol. 22(1), Springer-Verlag, Heidelberg (2004)

Hofmann, T.: Latent semantic models for collaborative filtering. ACM Trans. Inf. Syst. 22(1), 89–115 (2004)

Hu, Y., Koren, Y., Volinsky, C.: Collaborative filtering for implicit feedback datasets. In: IEEE International Conference on Data Mining (ICDM 2008), pp 263–272 (2008)

Huang, J., Guestrin, C., Guibas, L.: Efficient inference for distributions on permutations. In: Platt, J., Koller, D., Singer, Y., Roweis, S. (eds.) Advances in Neural Information Processing Systems, vol. 20, pp. 697–704. MIT Press, Cambridge (2008)

Kondor, R., Howard, A., Jebara, T.: Multi-object tracking with representations of the symmetric group. In: Proceedings of the Eleventh International Conference on Artificial Intelligence and Statistics, San Juan, Puerto Rico (2007)

Koren, Y.: Factorization meets the neighborhood: a multifaceted collaborative filtering model. In: KDD 2008: Proceeding of the 14th ACM SIGKDD International Conference on Knowledge Discovery and Data Mining, pp. 426–434. ACM, New York (2008)

Kurucz, M., Benczúr, A.A., Torma, B.: Methods for large scale svd with missing values. In: KDDCup 2007 (2007)

Linden, G., Smith, B., York, J.: Amazon.com recommendations: item-to-item collaborative filtering. Internet Computing 7(1), 76–80 (2003)

Marlin, B.: Modeling user rating profiles for collaborative filtering. In: Thrun, S., Saul, L., Schölkopf, B. (eds.) Advances in Neural Information Processing Systems, vol. 16, MIT Press, Cambridge (2004)

Pan, R., Zhou, Y., Cao, B., Liu, N.N., Lukose, R.M., Scholz, M., Yang, Q.: One-class collaborative filtering. In: IEEE International Conference on Data Mining (ICDM 2008), pp. 502–511 (2008)

Rendle, S., Schmidt-Thieme, L.: Online-updating regularized kernel matrix factorization models for large-scale recommender systems. In: RecSys 2008: Proceedings of the 2008 ACM Conference on Recommender Systems, pp. 251–258. ACM, New York (2008)

Rennie, J.D.M., Srebro, N.: Fast maximum margin matrix factorization for collaborative prediction. In: ICML 2005: Proceedings of the 22nd International Conference on Machine Learning, pp. 713–719. ACM, New York (2005)

Sarwar, B., Karypis, G., Konstan, J., Riedl, J.: Incremental singular value decomposition algorithms for highly scalable recommender systems. In: Proceedings of the 5th International Conference in Computers and Information Technology (2002)

Schmidt-Thieme, L.: Compound classification models for recommender systems. In: IEEE International Conference on Data Mining (ICDM 2005), pp. 378–385 (2005)

Srebro, N., Rennie, J.D.M., Jaakola, T.S.: Maximum-margin matrix factorization. In: Advances in Neural Information Processing Systems, vol. 17, pp. 1329–1336. MIT Press, Cambridge (2005)
Weimer, M., Karatzoglou, A., Smola, A.: Improving maximum margin matrix factorization. Machine Learning 72(3), 263–276 (2008)

Chapter 7
Tag Recommendation

Tagging is an important feature of the Web 2.0. It allows the user to annotate items/ resources like songs, pictures, bookmarks, etc. with keywords. Tagging helps the user to organize his items and facilitate e.g. browsing and searching. Tag recommenders assist the tagging process of a user by suggesting him a set of tags that he is likely to use for an item. Personalized tag recommenders take the user's tagging behaviour in the past into account when they recommend tags. That means each user is recommended a personalized list of tags – i.e. the suggested list of tags depends both on the user and the item. Personalization makes sense as people tend to use different tags for tagging the same item. This can be seen in systems like Last.fm that have a non-personalized tag recommender but still the people use different tags. For this, we will provide an empirical evaluation on a subset of Last.fm that shows, that our proposed personalized tag recommender outperform even the theoretical upper-bound for any non-personalized tag recommender.

In this work, we apply our context-aware ranking framework to tag recommendation. In tag recommendation, the context is the user/item-pair for which tags should be recommended. E.g. when a user wants to annotate a bookmark, the system provides recommendations that are both suitable for the bookmark and the user. That means in contrast to item recommendation where the context was just the user and the items should be ranked (see chapter 6), now tags/ keywords should be ranked for a context that includes both a user and an item. In total, tag recommendation is a three-mode problem.

First, we will discuss the related work in the area of tag recommenders. Then, the problem of tag recommendation is analyzed in detail and the relationship to context-aware ranking is shown. Afterwards, we show how the BCR method can be applied to tag recommendation. Here, we discuss both the optimization criterion BCR-OPT and the learning algorithm BCR-LEARN. The factorization models of chapter 5 can be used to represent the latent interactions within the data. For tag recommendation, the TD model has a cubic runtime in the factorization dimensionality (i.e. $O(k^3)$) and thus it is slow even for mid-sized dimensionalities. In contrast, the PITF and PARAFAC models have linear runtime. Learning TD with other optimization strategies than bootstrapped stochastic gradient descent can lead to faster learning.

S. Rendle: Context-Aware Ranking with Factorization Models, SCI 330, pp. 85–111.
springerlink.com

We will discuss therefore the relationships to Higher-Order-Singular-Value-Decomposition (HOSVD) for tag recommendation (Symeonidis et al, 2008) which corresponds to a dense least-square optimization. Furthermore, we describe post-wise AUC optimization of the Tucker Decomposition, which was introduced by us as *Ranking Tensor Factorization* (RTF) (Rendle et al, 2009). In the evaluation, we compare the PITF and PARAFAC model learned by BCR to RTF, HOSVD (=least-square TD) and the non-factorization approaches Folkrank and adapted Pagerank. We will show, that our tensor factorization approaches RTF and BCR-PITF provide the best quality improving both FolkRank and PageRank. Furthermore, we show that factorization approaches have a better prediction runtime that FolkRank. Finally, our experiments indicate that our method BCR-PITF outperforms the best quality method RTF-TD largely in runtime as the runtime drops from $O(k^3)$ to $O(k)$ — where k is the factorization dimension.

Besides lab experiments, our factorization models using the BCR-based optimization provided the best results for the ECML/PKDD Discovery Challenge 2009 for graph-based tag recommendation. This challenge was won by our BCR-PITF model (Rendle and Schmidt-Thieme, 2009).

7.1 Related Work

Even though tagging is a new trend in the WWW, tag recommendation has already attracted much research. Next, we will discuss this research and categorize it in personalized and non-personalized recommenders.

7.1.1 *Personalized Tag Recommendation*

Personalized tag recommendation is a recent topic in recommender systems. FolkRank, an adaption of PageRank, was introduced by Hotho et al (2006). FolkRank generates high quality recommendations (Jäschke et al, 2008) outperforming several baselines like most-popular models and collaborative filtering (Jäschke et al, 2007). Even though FolkRank showed to provide high quality recommendations, due to its very slow prediction runtime it is not applicable for large real-world scenarios. In contrast to this, the prediction runtime of factorization models are independent of the number of users and items – after the model has been learned.

Recently, factorization models based on Tucker Decomposition (TD) have been introduced to tag recommendation. Symeonidis et al (2008) apply a Higher-Order-Singular-Value-Decomposition (HOSVD) (Lathauwer et al, 2000) for computing a low rank approximation of the tensor of the observed data. This approach corresponds to a TD model optimized for square-loss where all not observed values are learned as 0s (see section 4.3.2). Thus, these dense square-loss approaches like HOSVD or other least-square methods (Lathauwer et al, 2000b) do not lead to optimal factorizations for the task of tag recommendation as we will show both theoretically and empirically.

Various methods have been proposed for tag recommendation at the ECML/ PKDD discovery challenge 2009 (DC09). We briefly describe the best performing methods. Marinho et al (2009) present a recommender method based on relational classification. A graph over posts is generated and tags are recommended based on the tags of the related neighbours. The paper also describes a semi-supervised extension that makes use of the posts that should be labeled in the future. For weighting the edges in the graph, several schemes based on cosine similarity are proposed. Zhang et al (2009) ensemble a collaborative filtering model based on user similarity with the FolkRank algorithm. The model in (Wetzker et al, 2009, 2010) is based on the assumption that different users have different vocabularies ('personomies'). Their approach is a linear combination between the most popular tags for an item and the estimated tags from the user's vocabulary to this item.

In contrast to the methods described so far, other approaches make use of additional content information. (Lipczak et al, 2009) propose an ensemble of six basic recommender which makes use of the text in the title and in URLs as well as resource related tags and the tags a user has given before. Similarly, Ju and Hwang (2009) extract candidates from textual content information of an item. They estimate the relevance of the candidates by their frequency and by matching them against the historical tags. This is combined with the tags the user has given in the past and the tags that were assigned to an item in the past. In section 7.6.3.1 our method is empirically compared these approaches on task 2 of the DC09.

7.1.2 Non-personalized Tag Recommendation

A non-personalized tag recommender predicts the same list of tags for the same item – i.e. it is independent of the user. There is several work on non-personalized tag recommenders, e.g. (Heymann et al, 2008; Song et al, 2008b,a). For example, Song et al (2008b) introduce an algorithm based on a Poisson mixture model. Although the algorithm is able to make predictions nearly in linear time, it is not personalized since the training data is composed from (words, documents, tags) triples containing no user specific information. Another difference to our work is that their method is content aware. Song et al (2008a) cast the problem of tag recommendation as a multi-label ranking problem for document classification and a fast recommendation algorithm based on gaussian processes is proposed. The algorithm provides linear time to train, proportional to the number of training samples, and constant time to predict per test case. Again differently from us, this approach is non-personalized since a given test document would be classified with the same set of tags independently of the users. Our evaluation (see section 7.6.3.1) indicates that if user information is present, our proposed personalized tag recommender outperforms any non-personalized tag recommender.

7.2 Personalized Tag Recommendation

Personalized tag recommendation is the task of recommending a list of tags to a user for annotating (e.g. describing) an item. An example is a music website where

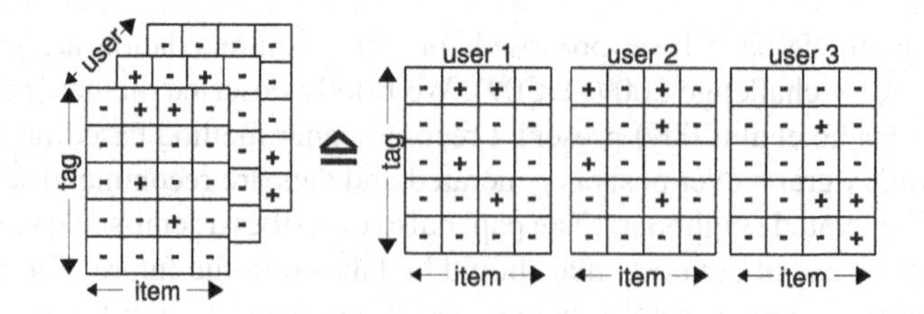

Fig. 7.1 The observed positive examples (u,i,t) are a ternary relationship that can be seen as a 3 dimensional tensor (cube). For each user a matrix is given that contains the tags given for a specific item.

a listener (user) wants to tag a song (item) and the system recommends him a list of keywords that the listener might want to use for this song. For inferring the recommendation list, a personalized tag recommender can use the historical data of the system, i.e. the tagging behaviour of the past. E.g. the recommender can make use of the tags that this user has given to other (similar) items in the past – or more general of similar tags that similar users haven given to similar items.

7.2.1 Formalization

Next, we show how tag recommendation can be expressed as a context-aware ranking problem of mode three ($m = 3$). For easier readability, we will use more meaningful names than X_i for the variable domains: $U = \{u_1, u_2, \ldots\}$ is the set of all users, $I = \{i_1, i_2, \ldots\}$ the set of all items and $T = \{t_1, t_2, \ldots\}$ the set of all tags. The historical tagging information of the past is given by $S \subseteq U \times I \times T$. As this is a ternary relation over categorical variables, it can be seen as a three-dimensional tensor (see figure 7.1) where the triples in S are the positive observations in the past. For tag recommendation, we are interested in recommending for a given user-item pair (u,i) a list of tags. That means, tag recommendation can be seen as context-aware ranking, where the context are the user-item pairs:

$$\mathscr{C} = U \times I \tag{7.1}$$

Following (Jäschke et al, 2007), we call each context $\mathbf{c} = (u,i) \in \mathscr{C}$ a *post* and we define the set of all observed posts P_S:

$$P_S := \{(u,i) | \exists t \in T : (u,i,t) \in S\}$$

P_S can be seen as a two-dimensional projection of S on the user/item dimension using the OR operation.

Now, the task of tag recommendation is the task of finding a context-aware ranking $\succ_{\mathbf{c}} = \succ_{u,i}$ for each post. Like suggested in section 3.4, all of the models presented here predict a scoring function $\hat{y} : U \times I \times T \to \mathbb{R}$ which can be used to derive a context-aware order according to eq. (3.34).

7.2.2 *Data Analysis*

The main problem in applying machine learning techniques on tagging system is that there are only observations S of positive tagging events (see figure 7.1). But it is unclear how the rest of this relation $(U \times I \times T) \setminus S$ should be interpreted.

7.2.2.1 0/1 Interpretation Scheme

A common interpretation scheme – we call it the *0/1 scheme* – is to encode positive feedback as 1 and interpret the remaining data as 0 (see figure 7.1). That means, the observed data is directly used for optimization. This interpretation is e.g. used for training tag recommenders using a HOSVD model (Symeonidis et al, 2008).

The 0/1 interpretation has three severe drawbacks:

1. The semantics are obviously incorrect. Imagine a user u has never tagged an item i before (e.g. figure 7.1, first item for user 1). For training a model with 0/1 interpretation all tags of this item are encoded with 0 and for learning the model is fitted to this data. So the model tries to predict a 0 for each case. The only reason why the model can predict something else than 0 is that it usually generalizes and does not fit exactly on the training data.
2. Also from a sparsity point of view the 0/1 scheme leads to a problem. If all elements that are not in S are assumed to be 0, even for a small dataset like Bibsonomy (see section 7.6.1), the 0 values dominate the 1 by many orders of magnitude. To give a practical example, first the sparsity for 0/1 interpretation is:

$$1 - \frac{|S|}{|U| \cdot |I| \cdot |T|} \tag{7.2}$$

   With this definition, for the BibSonomy 5-core dataset 99.94% elements are 0 and for the larger Last.fm 10-core dataset 99.998% are 0.
3. As one is interested in ranked lists, trying to fit to the numerical values of 1 and 0 is an unnecessary constraint. Instead only the qualitative difference between a positive and negative example is important. That means $\hat{y}$ of a positive example should be larger than that of a negative example.

7.2.2.2 Post-based Ranking Interpretation Scheme

Instead, we propose to infer pairwise ranking constraints D_S from S like we have discussed in section 3.3.1. The idea is that within a post (u,i), one can assume that a tag t_A is preferred over another tag t_B iff (u,i,t_A) has been observed and (u,i,t_B) has not been observed. An example is given in figure 7.2. In total, the training data D_S for pairwise constraints is defined as:

$$D_S := \{(u,i,t_A,t_B) : (u,i,t_A) \in S \wedge (u,i,t_B) \notin S\} \tag{7.3}$$

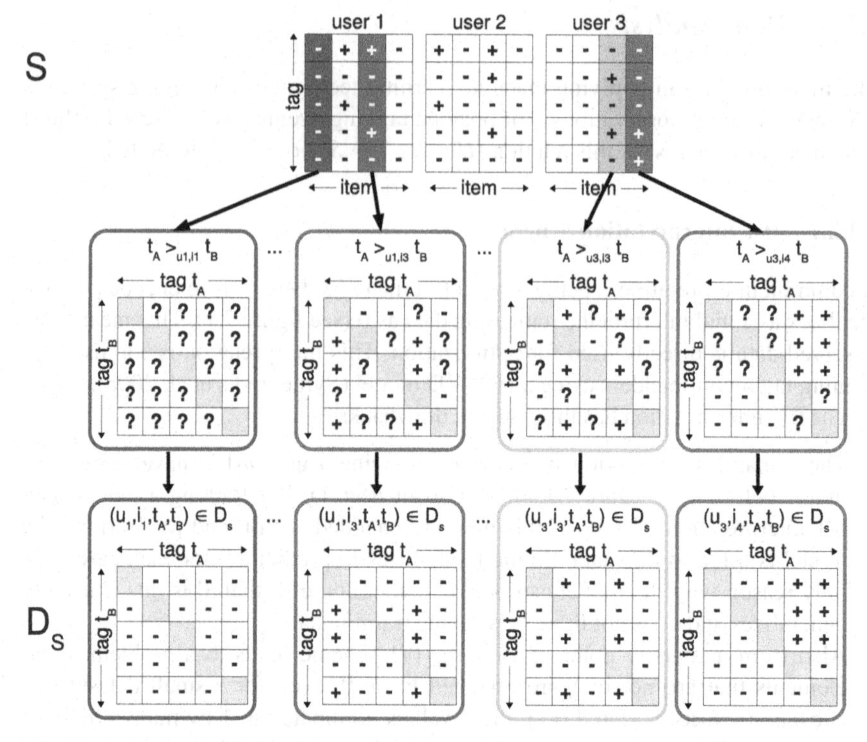

Fig. 7.2 Post-based ranking interpretation: From the observed data S, pairwise preferences D_S of tags can be inferred per post (user/ item combination). The figure shows examples for four posts: (u_1,i_1), (u_1,i_3), (u_3,i_3) and (u_3,i_4). E.g. for post (u_1,i_3), the following positive constraints can be inferred: $t_1 \succ_{u_1,i_3} t_2, t_1 \succ_{u_1,i_3} t_3, t_1 \succ_{u_1,i_3} t_5, t_4 \succ_{u_1,i_3} t_2, t_4 \succ_{u_1,i_3} t_3, t_4 \succ_{u_1,i_3} t_5$. For posts without any observed tag (like (u_1,i_1)), no constraints can be inferred.

The ranking relation $t_A \overset{?}{\succ}_{u,i} t_B$ between two non-observed tags $((u,i,t_A) \notin S \wedge (u,i,t_B) \notin S)$ is the one that should be predicted in the future. The main advantage of our approach is, that these pairs are treated as missing values (see the '?'s in figure 7.2, second row). Other approaches like (Symeonidis et al, 2008) learn that all these tags are not liked – i.e. they should have the same preference score 0. From a semantical point of view our scheme makes more sense as the user/ item combinations (posts) that have no tags (e.g. (u_1,i_1)), are the ones that the recommender system will have to predict a ranking for in the future. With our interpretation we treat this kind of data as missing values and do not use it as training data like in the '0/1 scheme'. Also inside a given post the non-observed tags are not fitted to 0, instead we only require that the positive examples have a higher value than the negative ones. This addresses the first two drawbacks of the '0/1 scheme'. The third drawback is tackled by our scheme by allowing free values for y and only posing pairwise ranking constraints (see eq. (3.34) and (7.3)). In all, a model for 'post-based ranking interpretation' should be optimized to satisfy as many ranking constraints as possible. In the following, we show how this can be done by the BCR method.

7.3 Bayesian Post-aware Ranking (BPoR) for Tag Recommendation

As we have shown in the last section, the problem of tag recommendation can be seen as an instance of context-aware ranking. Thus, the Bayesian Context-aware Ranking method gives the MAP estimator for the ranking $\succ$ that is defined by $\hat{y}$. Because the context is a post, we will refer to BCR for tag recommendation as *Bayesian Post-aware Ranking* (BPoR). Next, we show how to apply the optimization criterion BCR-OPT and the learning algorithm BCR-LEARN to tag recommenders.

*7.3.1 Optimization Criterion (*BPOR-OPT*)*

Like in section 4.1, the MAP estimator for the model parameters Θ that parametrize $\hat{y}$ is:

$$\underset{\Theta}{\operatorname{argmax}}\, p(\Theta \mid \succ) = \underset{\Theta}{\operatorname{argmax}}\, p(\succ \mid \Theta)\, p(\Theta) \tag{7.4}$$

where the probability for $\succ$ is defined over all context (i.e. combinations of users and items) and pairs of tags:

$$p(\succ \mid \Theta) = \prod_{(u,i,t_A,t_B)\in U\times I\times T^2} p(t_A \succ_{u,i} t_B \mid \Theta)^{2\,\delta(t_A \succ_{u,i} t_B)} \tag{7.5}$$

The probability of each quadruple (u,i,t_A,t_B) can be defined using the estimator $\hat{y}$ for $\succ$ (see section 4.1.3) and thus:

$$p(t_A \succ_{u,i} t_B|\Theta) := \sigma(\hat{y}(u,i,t_A) - \hat{y}(u,i,t_B)) \tag{7.6}$$

Again, gaussian priors are assumed over the model parameters:

$$p(\Theta) = \prod_{\theta\in\Theta} \sqrt{\frac{\lambda_\theta}{\pi}} \exp\left(-\lambda_\theta\, \theta^2\right) \tag{7.7}$$

Putting everything together, leads to BPoR-Opt for tag recomendation which is defined as the MAP estimator given the training data D_S:

$$\underset{\Theta}{\operatorname{argmax}}\, \text{BPOR-OPT} := \underset{\Theta}{\operatorname{argmax}} \sum_{(u,i,t_A,t_B)\in D_S} \ln \sigma(\hat{y}(u,i,t_A) - \hat{y}(u,i,t_B)) - \sum_{\theta\in\Theta} \frac{1}{2} \lambda_\theta\, \theta^2 \tag{7.8}$$

where λ_θ are model specific regularization parameters.

*7.3.2 Learning Algorithm (*BPOR-LEARN*)*

Secondly, we derive a learning algorithm to optimize the model parameters Θ of $\hat{y}$ for BPOR-OPT. In general, optimizing BPOR-OPT is time consuming, as D_S is very large – the size of D_S is in $O(|S|\,|T|)$. E.g. for the examples of our evaluation section this would be about 3,299,006,344 quadruples for the ECML/PKDD Discovery Challenge 09 and 449,290,590 quadruples for our Last.fm subset. Thus computing the full gradients is very slow and normal gradient descent is not feasible. Also stochastic gradient descent where the quadruples are traversed in a sorted way like per post or per user will be slow – an example for this can be found in figure 4.1. Instead, BPoR-LEARN draws quadruples by bootstrapping following the idea of BCR-LEARN (see section 4.2.2).

Gradients

The gradient of BPoR-OPT given a case (u,i,t_A,t_B) with respect to a model parameter θ is:

$$\frac{\partial}{\partial\theta}\text{BPoR-OPT} = \delta_{u,i,t_A,t_B}\frac{\partial}{\partial\theta}(\hat{y}(u,i,t_A) - \hat{y}(u,i,t_B)) - \lambda_\theta\,\theta \tag{7.9}$$

with:

$$\delta_{u,i,t_A,t_B} := (1 - \sigma(\hat{y}(u,i,t_A) - \hat{y}(u,i,t_B))) \tag{7.10}$$

That means, to apply BPoR-LEARN to a given model $\hat{y}$, only the gradient $\frac{\partial}{\partial\theta}\hat{y}(u,i,t)$ has to be known. In the next section, we derive our factorization models and also show their gradients for optimization w.r.t. BPoR-OPT using BPoR-LEARN.

Drawing of Cases

Like described before, the number of quadruples in D_S is huge. Thus, it is not feasible to explicitly enumerate all those quadruples. It is possible to draw cases (u,i,t_A,t_B) from D_S without enumerating them. The reason is, that D_S consists of all positive triples $(u,i,t_A) \in S$ combined with a negative example $(u,i,t_B) \notin S$. Therefore, first a triple (u,i,t_A) from S is drawn. From this a quadruple of D_S can be created by drawing a negative triple $(u,i,t_B) \notin S$. Such a negative triple can be drawn without enumerating them, by (1) drawing $t_B \in T$ and (2) rejecting it, if $(u,i,t_B) \in S$. This drawing scheme is effective because rejection is unlikely as most tags are not observed within a given post (u,i).

BPoR-Learn

Algorithm 3 shows the generic learning method BPoR-LEARN for optimizing BPoR-OPT for tag recommendation. Analogously to BCR-LEARN (algorithm 1) it first initializes the model parameters with random values. Then the parameters are learned iteratively by stochastic gradient descent. A case $(u,i,t_A,t_B) \in D_S$ is created using the drawing approach described above. Then the derivatives are computed and

Algorithm 3 BPOR-LEARN

Input: training data S, learning rate α, regularization parameters λ_θ
Output: model parameters Θ
1: initialize Θ from $\mathcal{N}(0, \sigma^2)$
2: **repeat**
3: draw (u, i, t_A) uniformly from S
4: draw t_B from $\{t : (u, i, t) \notin S\}$
5: $\delta_{u,i,t_A,t_B} \leftarrow (1 - \sigma(\hat{y}(u, i, t_A) - \hat{y}(u, i, t_B)))$
6: **for** $\theta \in \Theta$ **do**
7: $\theta \leftarrow \theta + \alpha \left(\delta_{u,i,t_A,t_B} \frac{\partial}{\partial \theta} (\hat{y}(u, i, t_A) - \hat{y}(u, i, t_B)) - \lambda_\theta \, \theta \right)$
8: **end for**
9: **until** convergence
10: **return** Θ

a small step of length α towards maximizing the quality is taken. This is repeated until a stopping criterion is met.

7.4 Factorization Models for Tag Recommendation

In the following, we apply three factorization models to tag recommendation: Tucker decomposition (TD), Parallel factor analysis (PARAFAC) and our pairwise interaction tensor factorization model (PITF) (see figure 7.3). We will show for each model how it can be learned with BPoR and the relationships to the other models. All of our factorization models predict a scoring function $\hat{y} : U \times I \times T \to \mathbb{R}$ which can be seen as a three-dimensional tensor Y where the value of entry (u, i, t) is the score $\hat{y}_{u,i,t}$.

7.4.1 Tucker Decomposition (TD)

Tucker decomposition (Tucker, 1966) factorizes a higher-order cube into a core tensor and one factor matrix for each dimensions.

$$\hat{y}^{\mathrm{TD}}_{u,i,t} := \sum_{f_U=1}^{k_U} \sum_{f_I=1}^{k_I} \sum_{f_T=1}^{k_T} b_{f_U,f_I,f_T} \cdot v^U_{u,f_U} \cdot v^I_{i,f_I} \cdot v^T_{t,f_T} \tag{7.11}$$

or equivalently as tensor product (see figure 7.3):

$$\hat{Y}^{\mathrm{TD}} := \mathcal{B} \times_U V^U \times_I V^I \times_T V^T \tag{7.12}$$

with model parameters:

$$V^U \in \mathbb{R}^{|U| \times k_U}, \quad V^I \in \mathbb{R}^{|I| \times k_I}, \quad V^T \in \mathbb{R}^{|T| \times k_T},$$
$$\mathcal{B} \in \mathbb{R}^{k_U \times k_I \times k_T}, \quad k_U, k_I, k_T \in \mathbb{N}^+ \tag{7.13}$$

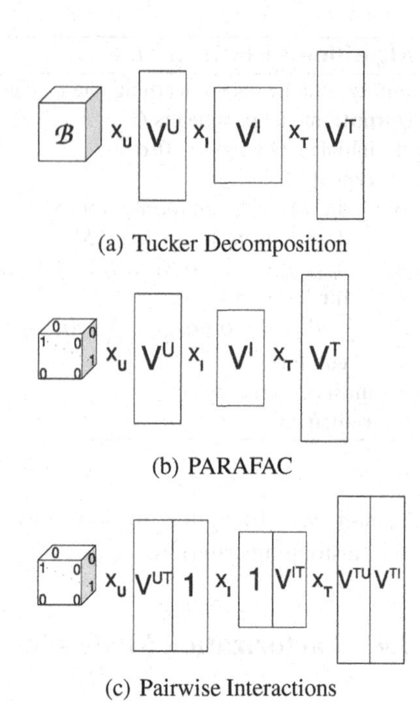

Fig. 7.3 Tensor Factorization models: $\mathcal{B}$, V^U, V^I and V^T are the model parameters (one tensor, three matrices). In Tucker Decomposition the core tensor $\mathcal{B}$ is variable and the factorization dimensions can differ. For PARAFAC and Pairwise Interactions the core is a fixed diagonal tensor. In the Pairwise Interaction model, parts of the feature matrices are fixed which corresponds to modelling pairwise interactions.

For learning such a TD model with BPOR-OPT, the partial derivatives of $\hat{y}^{\mathrm{TD}}$ given a case (u,i,t) are:

$$\frac{\partial \hat{y}^{\mathrm{TD}}_{u,i,t}}{\partial b_{f_U,f_I,f_T}} = v^U_{u,f_U} \cdot v^I_{i,f_I} \cdot v^T_{t,f_T} \tag{7.14}$$

$$\frac{\partial \hat{y}^{\mathrm{TD}}_{u,i,t}}{\partial v^U_{u,f_U}} = \sum_{f_I=1}^{k_I} \sum_{f_T=1}^{k_T} b_{f_U,f_I,f_T} \cdot v^I_{i,f_I} \cdot v^T_{t,f_T} \tag{7.15}$$

$$\frac{\partial \hat{y}^{\mathrm{TD}}_{u,i,t}}{\partial v^I_{i,f_I}} = \sum_{f_U=1}^{k_U} \sum_{f_T=1}^{k_T} b_{f_U,f_I,f_T} \cdot v^U_{u,f_U} \cdot v^T_{t,f_T} \tag{7.16}$$

$$\frac{\partial \hat{y}^{\mathrm{TD}}_{u,i,t}}{\partial v^T_{t,f_T}} = \sum_{f_U=1}^{k_U} \sum_{f_I=1}^{k_I} b_{f_U,f_I,f_T} \cdot v^U_{u,f_U} \cdot v^I_{i,f_I} \tag{7.17}$$

As discussed in section 5.1, the drawback of TD is that the model equation is a nested sum of degree 3 – i.e. it is cubic in $k := \min(k_u, k_i, k_t)$ and so the runtime complexity for predicting one triple (u,i,t) is in $O(k^3)$. Thus learning a TD model is slow even for a small to mid-sized number of factorization dimensions.

7.4.2 Parallel Factor Analysis (PARAFAC)

The PARAFAC model is a special case of the general Tucker decomposition model.

$$\hat{y}_{u,i,t}^{\text{PARAFAC}} := \sum_{f=1}^{k} v_{u,f}^{U} \cdot v_{i,f}^{I} \cdot v_{t,f}^{T} \tag{7.18}$$

As we have shown in section 5.2, PARAFAC can be derived from the Tucker decomposition model by setting $\mathscr{B}$ to the diagonal tensor.

The gradients of PARAFAC for tag recommendation are:

$$\frac{\partial \hat{y}_{u,i,t}^{\text{PARAFAC}}}{\partial v_{u,f}^{U}} = v_{i,f}^{I} \cdot v_{t,f}^{T}, \quad \frac{\partial \hat{y}_{u,i,t}^{\text{PARAFAC}}}{\partial v_{i,f}^{I}} = v_{u,f}^{U} \cdot v_{t,f}^{T}, \quad \frac{\partial \hat{y}_{u,i,t}^{\text{PARAFAC}}}{\partial v_{t,f}^{T}} = v_{u,f}^{U} \cdot v_{i,f}^{I} \tag{7.19}$$

Obviously, the PARAFAC model has a much better runtime complexity than TD because the model equation contains no nested sums and thus is in $O(k)$. A detailed discussion can be found in section 5.2.

7.4.3 Pairwise Interaction Tensor Factorization (PITF)

PITF explicitly models the two-way interactions between users, tags and items by factorizing each of the three relationships:

$$\hat{y}_{u,i,t}^{\text{PITF}} = \sum_{f=1}^{k_{U,T}} v_{u,f}^{U,T} \cdot v_{t,f}^{T,U} + \sum_{f=1}^{k_{I,T}} v_{i,f}^{I,T} \cdot v_{t,f}^{T,I} + \sum_{f=1}^{k_{U,I}} v_{u,f}^{U,I} \cdot v_{i,f}^{I,U} \tag{7.20}$$

The user-item interaction vanishes for predicting rankings and for BPoR optimization. The reason is that given a post (u,i), both the optimization criterion BPoR and the ranking ignores any score on the user-item interaction (see lemma 5.2). This results in the final model equation for PITF as follows:

$$\hat{y}_{u,i,t}^{\text{PITF}} = \sum_{f=1}^{k_{U,T}} v_{u,f}^{U,T} \cdot v_{t,f}^{T,U} + \sum_{f=1}^{k_{I,T}} v_{i,f}^{I,T} \cdot v_{t,f}^{T,I} \tag{7.21}$$

with model parameters:

$$V^{U,T} \in \mathbb{R}^{|U| \times k_{U,T}}, \quad V^{T,U} \in \mathbb{R}^{|T| \times k_{U,T}}, \quad k_{U,T} \in \mathbb{N}^{+} \tag{7.22}$$

$$V^{I,T} \in \mathbb{R}^{|I| \times k_{I,T}}, \quad V^{T,I} \in \mathbb{R}^{|T| \times k_{I,T}}, \quad k_{I,T} \in \mathbb{N}^{+} \tag{7.23}$$

As factorization dimensionality, we will always set $k = k_{U,T} = k_{I,T}$. In section 5.3, we have shown that PITF is a special case of PARAFAC and that the complexity of PITF is also in $O(k)$.

The gradients for the PITF model given a case (u,i,t) are:

$$\frac{\partial \hat{y}_{u,i,t}}{\partial v_{u,f}^{U,T}} = v_{t,f}^{T,U}, \quad \frac{\partial \hat{y}_{u,i,t}}{\partial v_{t,f}^{T,U}} = v_{u,f}^{U,T}, \quad \frac{\partial \hat{y}_{u,i,t}}{\partial v_{i,f}^{I,T}} = v_{t,f}^{T,I}, \quad \frac{\partial \hat{y}_{u,i,t}}{\partial v_{t,f}^{T,I}} = v_{i,f}^{I,T} \tag{7.24}$$

Algorithm 4 BPoR-LEARN-PITF

Input: training data S, learning rate α, regularization parameter λ
Output: model parameters $V^{U,T}, V^{T,U}, V^{I,T}, V^{T,I}$
1: initialize $V^{U,T}, V^{T,U}, V^{I,T}, V^{T,I}$ from $\mathcal{N}(0, \sigma^2)$
2: **repeat**
3: draw (u,i,t_A) uniformly from S
4: draw t_B from $\{t : (u,i,t) \notin S\}$
5: $\delta_{u,i,t_A,t_B} \leftarrow (1 - \sigma(\hat{y}(u,i,t_A) - \hat{y}(u,i,t_B)))$
6: **for** $f \in 1, \ldots, k$ **do**
7: $v^{U,T}_{u,f} \leftarrow v^{U,T}_{u,f} + \alpha \left(\delta_{u,i,t_A,t_B} \, (v^{T,U}_{t_A,f} - v^{T,U}_{t_B,f}) - \lambda \, \theta \right)$
8: $v^{I,T}_{i,f} \leftarrow v^{I,T}_{i,f} + \alpha \left(\delta_{u,i,t_A,t_B} \, (v^{T,I}_{t_A,f} - v^{T,I}_{t_B,f}) - \lambda \, \theta \right)$
9: $v^{T,U}_{t_A,f} \leftarrow v^{T,U}_{t_A,f} + \alpha \left(\delta_{u,i,t_A,t_B} \, v^{U,T}_{u,f} - \lambda \, \theta \right)$
10: $v^{T,U}_{t_B,f} \leftarrow v^{T,U}_{t_B,f} + \alpha \left(-\delta_{u,i,t_A,t_B} \, v^{U,T}_{u,f} - \lambda \, \theta \right)$
11: $v^{T,I}_{t_A,f} \leftarrow v^{T,I}_{t_A,f} + \alpha \left(\delta_{u,i,t_A,t_B} \, v^{I,T}_{i,f} - \lambda \, \theta \right)$
12: $v^{T,I}_{t_B,f} \leftarrow v^{T,I}_{t_B,f} + \alpha \left(-\delta_{u,i,t_A,t_B} \, v^{I,T}_{i,f} - \lambda \, \theta \right)$
13: **end for**
14: **until** convergence
15: **return** $V^{U,T}, V^{T,U}, V^{I,T}, V^{T,I}$

Algorithm 4 shows the complete BPoR optimization method for PITF.

7.4.4 Relation between TD, PARAFAC and PITF

In chapter 5, we have shown that for $m > 2$ (note in tag recommendation $m = 3$), the class of all PARAFAC models is a true subclass of TD and PITF is a true subclass of PARAFAC. I.e. here:

$$\mathcal{M}^{\text{TD}} \supset \mathcal{M}^{\text{PARAFAC}} \supset \mathcal{M}^{\text{PITF}} \tag{7.25}$$

From the perspective of expressiveness and complexity analysis, it does not make sense to use the PITF model. But in a sparse setting with little training data, it makes sense to use less expressive models with a fixed instead of a variable structure if this structure is carefully chosen. In our evaluation section, we empirically show that PITF models outperform PARAFAC models for tag recommendation when using Gaussian priors on the free model parameters like in eq. (7.7).

7.5 Alternative Optimization for Tucker Decomposition

Next, we want to discuss two alternative optimization criteria for TD models. The first one is HOSVD which corresponds to a dense least-square optimization of the TD parameters. The second one is RTF, a ranking optimization based on post-wise AUC maximization.

7.5.1 *Higher-Order Singular Value Decomposition (HOSVD)*

Singular Value Decomposition (SVD) is a well studied factorization method for two mode problems. It creates the best k-rank approximation of a matrix with respect to minimal least-square error. Higher-order singular value decomposition (HOSVD) is an extension of SVD to problems of higher mode ($m \geq 3$). The model equation of HOSVD is the Tucker decomposition model. But using (HO)SVD always implies that the optimization criterion is least square. Moreover (HO)SVD does not handle any missing values, i.e. the least square of all elements in the tensor is calculated. The optimization criterion for HOSVD for tag recommendation is:

$$\underset{\mathscr{B},V^U,V^I,V^T}{\operatorname{argmin}} \sum_{(u,i,t)\in(U\times I\times T)} (\hat{y}_{u,i,t} - \delta((u,i,t)\in S))^2 \tag{7.26}$$

This corresponds to the 0/1 interpretation scheme, that we have discussed in section 7.2.2.1.

An advantage of HOSVD is that there are fast optimization algorithms for sparse settings – here sparsity means many 0 values. Lathauwer et al (2000) introduce an approximation that first generates three two-mode problems by unfolding the tensor. Then each two-mode problem is solved with a sparse SVD solver. The factorization matrices of the two-mode problems can be used as factorization matrices for the original problem. Finally, the core tensor can be calculated from the factorization matrices and the data.

In our evaluation, we will compare this approach to our proposed BPoR optimization.

7.5.2 *Optimizing the Ranking Statistic AUC per Post (RTF)*

In (Rendle et al, 2009), we have developed an improvement of HOSVD for the task of tag recommendation that optimizes the AUC on the observed posts. The method was named RTF for 'Ranking with Tensor Factorization'. In the following, we will describe this method and show the similarities and differences to BCR.

7.5.2.1 Optimization Criterion

RTF uses the same pairwise data interpretation D_S as in eq. (7.3). But the optimization is done with respect to the AUC over each observed post instead of BCR:

$$\underset{\mathscr{B},V^U,V^I,V^T}{\operatorname{argmax}} \sum_{(u,i)\in P_S} \operatorname{AUC}(u,i) \tag{7.27}$$

with

$$\operatorname{AUC}(u,i) := \frac{1}{|T^+_{u,i}||T^-_{u,i}|} \sum_{t_A\in T^+_{u,i}} \sum_{t_B\in T^-_{u,i}} \delta(\hat{y}_{u,i,t_A} > \hat{y}_{u,i,t_B}) \tag{7.28}$$

where:

$$T_{u,i}^{+} := \{t : (u,i,t) \in S\}, \quad T_{u,i}^{-} := T \setminus T_{u,i}^{+} \tag{7.29}$$

In section 4.3.1, we have shown that eq. (7.27) can be rewritten as:

$$\underset{\mathscr{B},V^U,V^I,V^T}{\operatorname{argmax}} \sum_{(u,i,t_A,t_B)\in D_S} z_{u,i}\, \delta(\hat{y}_{u,i,t_A} > \hat{y}_{u,i,t_B}) \tag{7.30}$$

where $z_{u,i}$ is the normalization constant:

$$z_{u,i} = \frac{1}{|T_{u,i}^{+}||T_{u,i}^{-}|} \tag{7.31}$$

Regularization

The optimization criterion presented so far will lead to the best value given the training data. With high feature dimensions (i.e. high k_U, k_I, k_T) an arbitrary small error on the training data can be achieved. In general we are not interested in a low error for the already observed data but in a low error over unseen data. Minimizing the training error for models with a large number of parameters will lead to overfitting, i.e. a small training error but a large error over new/ unseen data. A common way to prevent this is to regularize the optimization criterion. Regularization is very successful in related areas like rating prediction (Rennie and Srebro, 2005). Adding a regularization objective to the optimization task in formula (7.27) leads to the following objective:

$$\underset{\mathscr{B},V^U,V^I,V^T}{\operatorname{argmax}} \sum_{(u,i)\in P_S} \operatorname{AUC}(u,i) - \frac{1}{2}\left(\lambda_{\mathscr{B}}||\mathscr{B}||_F^2 + \lambda_U||V^U||_F^2 + \lambda_I||V^I||_F^2 + \lambda_T||V^T||_F^2\right) \tag{7.32}$$

Where $\lambda_U, \lambda_I, \lambda_T, \lambda_{\mathscr{B}} \in \mathbb{R}_0^+$ are the regularization parameters for the the feature matrices and core tensor respectively.

7.5.2.2 Learning Algorithm

For optimizing eq. (7.27), the δ-function is approximated by the differentiable σ-function. As the AUC per observed post $(u,i) \in P_S$ should be optimized, we perform gradient descent per post, i.e. $\frac{\partial}{\partial\theta}\operatorname{AUC}(u,i)$ (see algorithm 5).

Gradients

The gradients for each model parameter given a post (u,i) are:

Algorithm 5 RTF-LEARN

Input: training data S, learning rate α, regularization parameters $\lambda_{\mathscr{B}}, \lambda_U, \lambda_I, \lambda_T$
Output: model parameters $\mathscr{B}, V^U, V^I, V^T$

1: initialize $\mathscr{B}, V^U, V^I, V^T$ from $\mathscr{N}(0, \sigma^2)$
2: **repeat**
3: **for** $(u,i) \in P_S$ **do**
4: **for** $(f_U, f_I, f_T) \in k_U \times k_I \times k_T$ **do**
5: $b_{f_U,f_I,f_T} \leftarrow b_{f_U,f_I,f_T} + \alpha \left(\frac{\partial}{\partial b_{f_U,f_I,b_T}} \text{AUC}(u,i) - \lambda_{\mathscr{B}} b_{f_U,f_I,f_T} \right)$
6: **end for**
7: **for** $f_U \in \{1, \ldots, k_U\}$ **do**
8: $v^U_{u,f_U} \leftarrow v^U_{u,f_U} + \alpha \left(\frac{\partial}{\partial v^U_{u,f_U}} \text{AUC}(u,i) - \lambda_U v^U_{u,f_U} \right)$
9: **end for**
10: **for** $f_I \in \{1, \ldots, k_I\}$ **do**
11: $v^I_{i,f_I} \leftarrow v^I_{i,f_I} + \alpha \left(\frac{\partial}{\partial v^I_{i,f_I}} \text{AUC}(u,i) - \lambda_I v^I_{i,f_I} \right)$
12: **end for**
13: **for** $t \in T$ **do**
14: **for** $f_T \in \{1, \ldots, k_T\}$ **do**
15: $v^T_{t,f_T} \leftarrow v^T_{t,f_T} + \alpha \left(\frac{\partial}{\partial v^T_{t,f_T}} \text{AUC}(u,i) - \lambda_T v^T_{t,f_T} \right)$
16: **end for**
17: **end for**
18: **end for**
19: **until** convergence
20: **return** $\mathscr{B}, V^U, V^I, V^T$

$$\frac{\partial}{\partial b_{f_U,f_I,f_T}} \text{AUC}(u,i) = z_{u,i} \sum_{t_A \in T^+_{u,i}} \sum_{t_B \in T^-_{u,i}} \delta_{u,i,t_A,t_B} \, v^U_{u,f_U} \, v^I_{i,f_I} \, (v^T_{t_A,f_T} - v^T_{t_B,f_T}) \tag{7.33}$$

$$\frac{\partial}{\partial v^U_{u,f_U}} \text{AUC}(u,i) = z_{u,i} \sum_{t_A \in T^+_{u,i}} \sum_{t_B \in T^-_{u,i}} \sum_{f_I=1}^{k_I} \sum_{f_T=1}^{k_T} \delta_{u,i,t_A,t_B} \, b_{f_U,f_I,f_T} \, v^I_{i,f_I} \, (v^T_{t_A,f_T} - v^T_{t_B,f_T}) \tag{7.34}$$

$$\frac{\partial}{\partial v^I_{i,f_I}} \text{AUC}(u,i) = z_{u,i} \sum_{t_A \in T^+_{u,i}} \sum_{t_B \in T^-_{u,i}} \sum_{f_U=1}^{k_U} \sum_{f_T=1}^{k_T} \delta_{u,i,t_A,t_B} \, b_{f_U,f_I,f_T} \, v^U_{u,f_U} \, (v^T_{t_A,f_T} - v^T_{t_B,f_T}) \tag{7.35}$$

with:

$$\delta_{u,i,t_A,t_B} := \sigma(\hat{y}_{u,i,t_A} - \hat{y}_{u,i,t_B}) \, (1 - \sigma(\hat{y}_{u,i,t_A} - \hat{y}_{u,i,t_B})) \tag{7.36}$$

The gradients for the tags depend on whether the tag t is observed $t = t_A \in T^+_{u,i}$ or not $t = t_B \in T^-_{u,i}$:

$$\frac{\partial}{\partial v^T_{t_A,f_I}} \mathrm{AUC}(u,i) = z_{u,i} \sum_{t_B \in T^-_{u,i}} \sum_{f_U=1}^{k_U} \sum_{f_I=1}^{k_I} \delta_{u,i,t_A,t_B}\, b_{f_U,f_I,f_T}\, v^U_{u,f_U}\, v^I_{i,f_I} \tag{7.37}$$

$$\frac{\partial}{\partial v^T_{t_B,f_I}} \mathrm{AUC}(u,i) = -z_{u,i} \sum_{t_A \in T^+_{u,i}} \sum_{f_U=1}^{k_U} \sum_{f_I=1}^{k_I} \delta_{u,i,t_A,t_B}\, b_{f_U,f_I,f_T}\, v^U_{u,f_U}\, v^I_{i,f_I} \tag{7.38}$$

Fast Computation of Gradients

By storing intermediate results, these gradients can be calculated more efficiently. For the updates on the user, item and core, we precalculate the vector $\gamma \in \mathbb{R}^{k_T}$:

$$\gamma_{f_T} := \sum_{t^+ \in T^+_{u,i}} \sum_{t^- \in T^-_{u,i}} \delta_{u,i,t_A,t_B} (v^T_{t_A,f_T} - v^T_{t_B,f_T}) \tag{7.39}$$

The computation of the whole vector γ can be performed in $O(k_T\,|T^+|\,|T^-|)$. Now the gradients for the user, item and the core simplify to:

$$\frac{\partial}{\partial b_{f_U,f_I,f_T}} \mathrm{AUC}(u,i) = z_{u,i} \gamma_{f_T}\, v^U_{u,f_U}\, v^I_{i,f_I} \tag{7.40}$$

$$\frac{\partial}{\partial v^U_{u,f_U}} \mathrm{AUC}(u,i) = z_{u,i} \sum_{f_I=1}^{k_I} \sum_{f_T=1}^{k_T} \gamma_{f_T}\, b_{f_U,f_I,f_T}\, v^I_{i,f_I} \tag{7.41}$$

$$\frac{\partial}{\partial v^I_{i,f_I}} \mathrm{AUC}(u,i) = z_{u,i} \sum_{f_U=1}^{k_u} \sum_{f_T=1}^{k_T} \gamma_{f_T}\, b_{f_U,f_I,f_T}\, v^U_{u,f_U} \tag{7.42}$$

$$\tag{7.43}$$

With this, updating the whole core tensor ($k_U \cdot k_I \cdot k_T$ factors) and updating the k_U factors of user u and k_I factors of item i is in $O(k_T\,|T^+|\,|T^-| + k_U\,k_I\,k_T)$.

Similarly, for the tags the update can be simplified by storing the vector $\eta \in \mathbb{R}^{k_T}$:

$$\eta_{f_T} := \sum_{f_U=1}^{k_U} \sum_{f_I=1}^{k_I} b_{f_U,f_I,f_T}\, v^U_{u,f_U}\, v^U_{u,f_U} \tag{7.44}$$

which can be computed for the whole vector in $O(k_U\,k_I\,k_T)$. Now the gradients for the tags simplify to:

$$\frac{\partial}{\partial v^T_{t_A,f_I}} \mathrm{AUC}(u,i) = z_{u,i} \sum_{t_B \in T^-_{u,i}} \delta_{u,i,t_A,t_B}\, \eta_{f_T} \tag{7.45}$$

$$\frac{\partial}{\partial v^T_{t_B,f_I}} \mathrm{AUC}(u,i) = -z_{u,i} \sum_{t_A \in T^+_{u,i}} \delta_{u,i,t_A,t_B}\, \eta_{f_T} \tag{7.46}$$

To update the k_T factors for each of the $|T^+_{u,i}|$ tags that are observed, the complexity is $O(k_U\, k_I\, k_T + |T^+|\, k_T\, |T^-|)$. Similarly, for the $|T^+_{u,i}|$ negative tags, the complexity is $O(k_U\, k_I\, k_T + |T^-|\, k_T\, |T^+|)$.

In total, a gradient descent step for a whole post can be performed in $O(k_U\, k_I\, k_T + k_T\, |T^+||T^-|)$. Mostly, the number of tags given $|T^+_{u,i}|$ is constant and independent of $|T|$ because the tags a user gives does not depend on how many tags (words) there are in total. Furthermore, $|T^+_{u,i}|$ is typically small. Under these assumptions, the complexity is in $O(k_U\, k_I\, k_T + k_T\, |T|)$.

Fast Prediction with TD Models

For predicting $\hat{y}_{u,i,t}$ with a TD model, formula (7.11) is used. The runtime complexity for eq. (7.11) is $O(k_U \cdot k_I \cdot k_T)$ and thus the trivial upper bound of the runtime for predicting a top-n list is $O(|T| \cdot k_U \cdot k_I \cdot k_T)$. Though the runtime can be improved largely by reordering the sums in eq. (7.11):

$$\begin{aligned}
\hat{y}_{u,i,t} &= \sum_{f_U=1}^{k_U} \sum_{f_I=1}^{k_I} \sum_{f_T=1}^{k_T} b_{f_U,f_I,f_T}\, v^U_{u,f_U}\, v^I_{i,f_I}\, v^T_{t,f_T} \\
&= \sum_{f_T=1}^{k_T} v^T_{t,f_T} \sum_{f_U=1}^{k_U} \sum_{f_I=1}^{k_I} b_{f_U,f_I,f_T}\, v^U_{u,f_U}\, v^I_{i,f_I} \\
&= \sum_{f_T=1}^{k_T} v^T_{t,f_T}\, \rho_{f_T}
\end{aligned} \tag{7.47}$$

with:

$$\rho_{f_T} := \sum_{f_U=1}^{k_U} \sum_{f_I=1}^{k_I} b_{f_U,f_I,f_T}\, v^U_{u,f_U}\, v^I_{i,f_I} \tag{7.48}$$

When making a top-n prediction for user u and item i instead of computing (7.11) for each tag, the intermediate result vector ρ can be computed first in $O(k_U \cdot k_I \cdot k_T)$. The top-n prediction can then be made using this intermediate result and the total runtime of predicting top-n is then $O(|T| \cdot k_T + k_U \cdot k_I \cdot k_T)$. Thus the runtime for prediction with the TD model is independent of the number of users, items and observations S. It only depends on the dimensions of the factorization and the number of tags.

7.5.2.3 Comparison to BPoR

As we have described in section 4.3.1, the optimization criterion of AUC is related to BPoR. The difference is the normalization constant $z_{u,i}$ of posts and that as error measure, σ (for AUC) instead of $\ln \sigma$ (for BPoR) is used. But RTF also differs from BPoR in terms of the learning algorithm. For RTF, the post-wise AUC is directly optimized by performing gradient descent on posts (u,i) whereas for BPoR quadruples (u,i,t_A,t_B) are drawn. We will compare the empirical quality of BPoR, RTF and HOSVD next.

7.6 Evaluation

In our evaluation, we investigate the learning runtime and prediction quality of our proposed methods. For the runtime, we want to justify the results of the theoretical complexity analysis (TD is in O(k^3), PARAFAC/PITF in O(k)) by an empirical comparison of the TD model to the PARAFAC model and our PITF model. With respect to prediction quality, we investigate empirically whether the speedup of PARAFAC/ PITF is paid with quality – i.e. if there is a trade-off between quality and runtime between the model classes.

7.6.1 Datasets

We use three datasets for evaluation: Bibsonomy and Last.fm like in (Jäschke et al, 2007; Rendle et al, 2009) and the dataset from the ECML/ PKDD Discovery Challenge 2009[1]. All datasets are p-cores[2] – for BibSonomy the 5-core, for Last.fm the 10-core and for the ECML/PKDD Challenge the provided 2-core. The characteristics of the datasets can be found in table 7.1.

Table 7.1 Dataset characteristics in terms of number of users, items, tags, tagging triples S and posts.

Dataset	Users $\|U\|$	Items $\|I\|$	Tags $\|T\|$	Triples $\|S\|$	Posts $\|P_S\|$
BibSonomy	116	361	412	10,148	2,522
Last.fm	2,917	1,853	2,045	219,702	75,565
ECML/PKDD DC 09	1,185	22,389	13,276	248,494	63,628

7.6.2 Evaluation Methodology

For Bibsonomy and Last.fm we use the same protocol as described in (Jäschke et al, 2008) – i.e. per user one post is randomly removed from the training set S_{train} and put into the test set S_{test}. We use exactly the same splits as in (Jäschke et al, 2008). After the splits have been built, the recommenders are trained on the test set and then the prediction quality on the test set is measured. As quality measure, we report the F-Measure on top-N lists (see section 3.5). The experiments are repeated 10 times by sampling new training/ test sets. We report the average over all runs.

For the ECML Challenge dataset, we use the protocol and split of the challenge and report the official results.

[1] `http://www.kde.cs.uni-kassel.de/ws/dc09`

[2] The p-core of S is the largest subset of S with the property that every user, every item and every tag has to occur at least p times.

Hyperparameters

The hyperparameters of all models are searched on the first training split. For the RTF-TD model, the hyperparameters are: learning rate $\alpha = 0.5$ for BibSonomy and $\alpha = 0.1$ for Last.fm; regularization $\gamma = \gamma_c = 10^{-5}$ for BibSonomy and $\gamma = \gamma_c = 10^{-6}$ for Last.fm; iterations $iter = 500$ for BibSonomy and $iter = 600$ for Last.fm. The model parameters Θ are initialized with small random values drawn from the normal distribution $N(0, 0.1)$. For HOSVD we have a dimensionality of $(k_U, k_I, k_T) = (60, 105, 225)$ for BibSonomy and $(k_U, k_I, k_T) = (875, 556, 614)$ for Last.fm. For PITF the hyperparameters are $\lambda = 5e - 05$ and $\alpha = 0.05$. For PARAFAC they are $\lambda = 0$ and $\alpha = 0.01$. The parameters of both PITF and PARAFAC are initialized with $N(0, 0.01)$. For FolkRank and PageRank, we report the values obtained by Jäschke et al (2008) as we use the same datasets and splits.

Implementations

The learning runtime measurements of RTF-TD, BPoR-PITF and BPoR-PARAFAC were made with C++ implementations. The runtime measurement for predicting is made with a Object Pascal implementation of RTF-TD and a C++ implementation of Folkrank. In general, the experiments were run on a compute cluster with 200 cores in total. Each compute node has identical hard- and software. Our implementations use no parallelization neither over compute nodes nor within nodes – i.e. per run only one processor core was used.

7.6.3 Results

We compare our factorization models BPoR-PITF, BPoR-PARAFAC and RTF-TD to other state-of-the-art personalized tag recommender and the upper bound for non-personalized tag-recommender. We investigate both the quality of the predictions and the runtime for learning and predicting.

7.6.3.1 Prediction Quality

First of all, we compare the prediction quality of our factorization models BPoR-PITF, BPoR-PARAFAC and RTF-TD to competing models. In figure 7.4, a comparison to FolkRank, PageRank and HOSVD on Bibsonomy and Last.fm is shown. In general, the factorization models result in the best prediction quality – only on the very small Bibsonomy dataset FolkRank is competitive.

PITF vs. PARAFAC

When comparing the two factorization models with linear runtime in k – i.e. PARAFAC and PITF – one can see that BPoR-PITF achieves on all datasets a higher

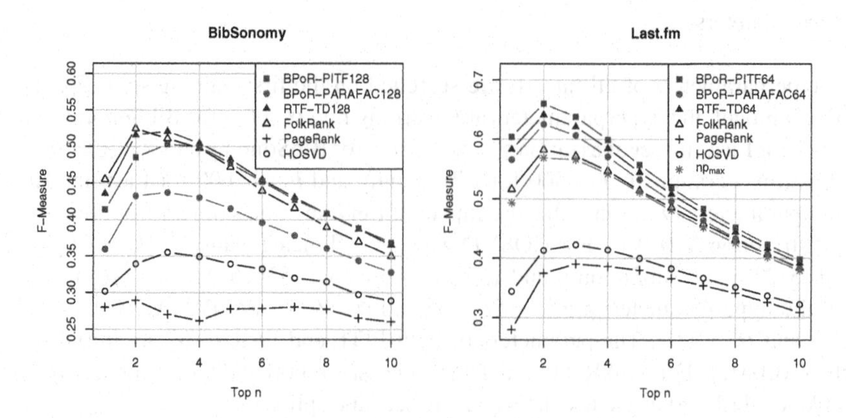

Fig. 7.4 Our factorization models (RTF-TD, BPoR-PARAFAC, BPoR-PITF) achieve the best prediction quality outperforming other approaches like FolkRank, PageRank and HOSVD. On the larger Last.fm dataset the BPoR-PITF model has the highest quality.

prediction quality than BPoR-PARAFAC. At first, this might be surprising because PARAFAC is more general and includes PITF. But it seems that BPoR-PARAFAC is unable to find the pairwise structure of PITF and to do regularization at the same time. An indication for this is that for PARAFAC the 'best' regularization parameter found by grid search is $\lambda = 0$.

PITF vs. RTF-TD

Next, we compare the prediction quality of the pairwise interaction model (BPoR-PITF) to full Tucker decomposition (RTF-TD) (see figure 7.4 and 7.5). On the small Bibsonomy dataset, on small top-N lists (1,2,3) RTF-TD outperforms BPoR-PITF whereas on larger lists, the difference vanishes. In contrast to this on the larger Last.fm dataset BPoR-PITF outperforms RTF-TD on all list sizes. These results indicate that the learning speedup of BPoR-PITF models to RTF-TD does not come to the prize of lower prediction quality. Rather, BPoR-PITF can even outperform RTF-TD in quality on larger datasets.

RTF-TD vs. HOSVD

The prediction quality of RTF-TD is clearly superior to the one of HOSVD. On BibSonomy even with a very small number of 8 dimensions, RTF-TD achieves almost similar results as HOSVD with a dimensionality of $(60, 105, 225)$ and $(875, 556, 614)$ for Last.fm respectively. Increasing the dimensions of RTF to 16 dimensions already largely outperforms HOSVD in quality. Note that for Last.fm this means that for HOSVD there are $298,711,000$ parameters to learn in the core tensor – whereas for RTF8 there are only 512 and for RTF16 only $4,096$ parameters.

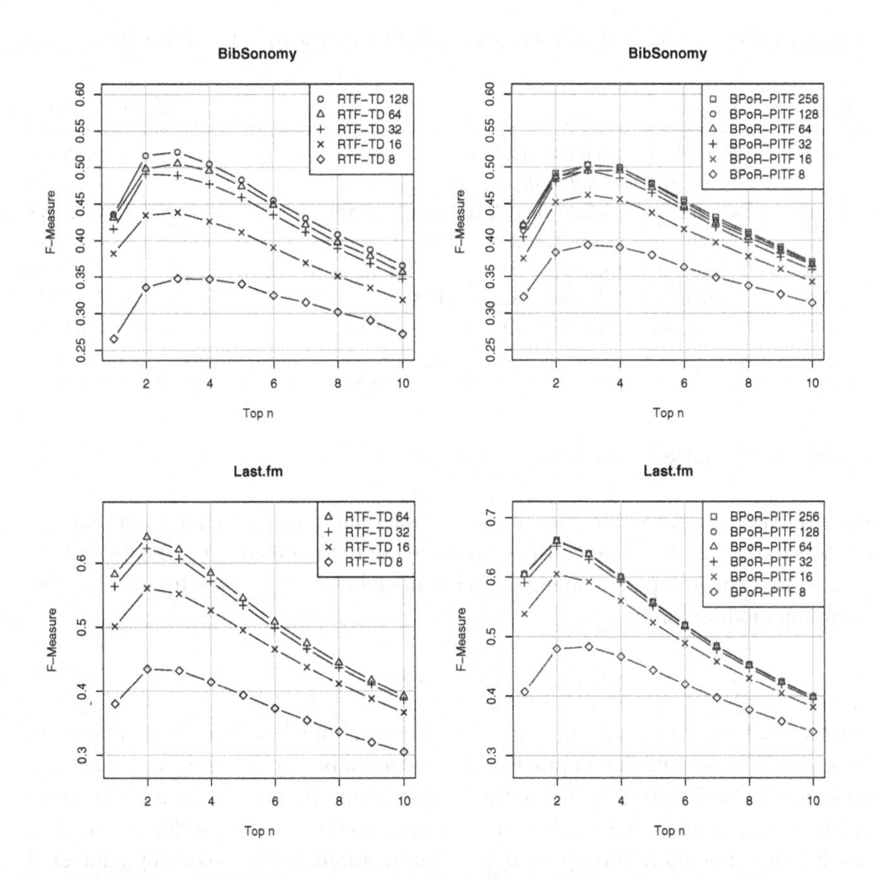

Fig. 7.5 PITF vs. RTF-TD for a varying number of factorization dimensions (k).

The empirical qualitative results match our discussion about the data interpretation in section 7.2.2.1.

Even though RTF-TD and HOSVD have the same prediction method (the Tucker decomposition) and thus prediction complexity, in practice RTF-TD models are much faster in prediction than comparable HOSVD models, because RTF-TD models need much less dimensions than HOSVD for achieving better quality. A final problem with HOSVD is that we found it to be very sensitive for the number of dimensions and that they have to be chosen carefully. Also HOSVD is sensitive to the relations between the user, item and tag dimensions – e.g. choosing the same dimension for all three dimensions leads to poor results. In contrast to this, for RTF-TD we can choose the same number of dimensions for user, item and tags. Furthermore for RTF-TD, by increasing the number of dimensions we get better results. We expect this behaviour due to the regularization of RTF-TD models.

Table 7.2 Official results (top-6) from the ECML/ PKDD Discovery Challenge 2009.

Rank	Method	Top-5 F-Measure
1	BPoR-PITF + adaptive list size	0.35594
-	*BPoR-PITF (not submitted)*	*0.345*
2	Relational Classification (Marinho et al, 2009)	0.33185
3	Content-based (Lipczak et al, 2009)	0.32461
4	Collaborative Filtering + FolkRank (Zhang et al, 2009)	0.32230
5	Content-based (Ju and Hwang, 2009)	0.32134
6	Personomy translation (Wetzker et al, 2009)	0.32124
...	...	...

Non-personalized Recommenders

In a last experiment, we compare the prediction quality of personalized tag recommenders to the best possible non-personalized tag recommender, i.e. the theoretical upper bound for **n**on-**p**ersonalized tag recommendation $\mathrm{np}_{\mathrm{max}}$ (see figure 7.4). The weighting method for $\mathrm{np}_{\mathrm{max}}$ is:

$$\hat{y}_{u,i,t}^{\mathrm{np}_{\mathrm{max}}} := |\{u'|(u',i,t) \in S_{test}\}| \tag{7.49}$$

That means, for each item the tags are ranked by counting how often a tag appears in the *test* set. Please note that in practice $\hat{y}^{\mathrm{np}_{\mathrm{max}}}$ cannot be applied as S_{test} is unknown. But here we use $\hat{y}^{\mathrm{np}_{\mathrm{max}}}$ as the theoretical upper bound for non-personalized recommenders because it creates the best non-personalized top-n list for the test set S_{test} – every other method for non-personalized tag recommendation like Heymann et al (2008); Song et al (2008b,a) is guaranteed to have a lower (or in the best case the same) quality on S_{test}. As figure 7.4 shows, personalized tag recommenders like FolkRank, PITF and RTF outperform $\mathrm{np}_{\mathrm{max}}$ the theoretical upper bound for non-personalized tag recommendation[3]. That means, in applications where personalized information is present, personalized tag recommenders are supposed to outperform non-personalized tag recommenders.

ECML / PKDD Discovery Challenge 09

In addition to the lab experiments, our BPoR-PITF model took also part in task 2 of the ECML/PKDD Discovery Challenge 09 and achieved the highest prediction quality. Table 7.2 shows the final results[4] listing the first six approaches. This evaluation in a tag recommender challenge organized by a third party shows that BPoR-PITF is able to create high quality predictions.

[3] Evaluating $\mathrm{np}_{\mathrm{max}}$ on the small BibSonomy dataset makes no sense because in the test sets S_{test} of BibSonomy are rarely two posts with the same item.

[4] `http://www.kde.cs.uni-kassel.de/ws/dc09/results`

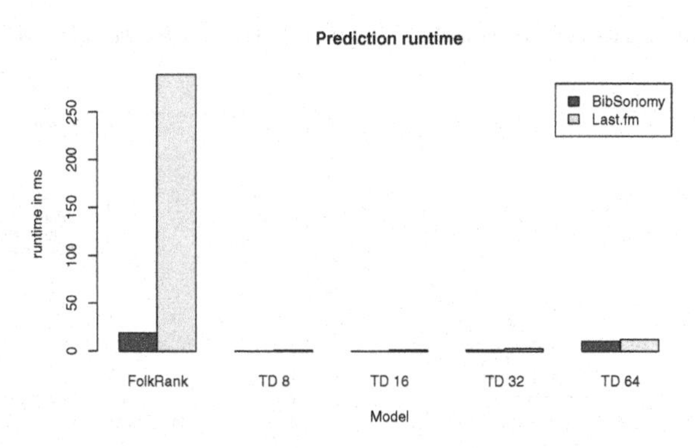

Fig. 7.6 Runtime comparison for predicting one ranked list of tags for the small BibSonomy and the larger Last.fm dataset. FolkRank is compared to a Tucker decomposition (e.g. RTF, HOSVD) with an increasing number of dimensions. On small datasets FolkRank's runtime is feasible but on larger datasets it gets impractical. In contrast to this, Tucker factorization models only depend on the factorization dimensions and not on the size of the dataset.

Our approach at the ECML/PKDD Challenge (Rendle and Schmidt-Thieme, 2009) had two additions to the BPoR-PITF presented here: (1) In the challenge, the recommender could benefit from suggesting lists with less than 5 tags – thus we estimated how many tags to recommend. Even without this enhancement for the challenge, our approach would still have the best score with 0.345. (2) We ensembled many BPoR-PITF models to reduce variance in the ranking estimates. On our holdout test, this improved the result only a little bit.

7.6.3.2 Runtime

Our results for the prediction quality indicate that BPoR-PITF and RTF-TD outperform other approaches. Next, we want to investigate the runtime for prediction and learning.

Prediction runtime

Fast predictions are crucial for applying tag recommenders. We compare the factorization model with the largest prediction complexity (i.e. Tucker decomposition) to Folkrank which achieves the best quality among the non-factorization approaches.

The empirical runtime comparison for predicting a ranked list of tags for a post can be found in figure 7.6. As you can see, the runtime of a TD model[5] is dominated by the dimension of the factorization and is more or less independent of the size of the dataset. The runtime on the BibSonomy dataset and the 20 times larger Last.fm dataset are almost the same – e.g. for $k = 64$ 10.4 ms for BibSonomy and 12.4 ms

[5] Note that the prediction runtime complexity of a Tucker decomposition model is independent of the optimization (e.g. RTF, BPoR, HOSVD).

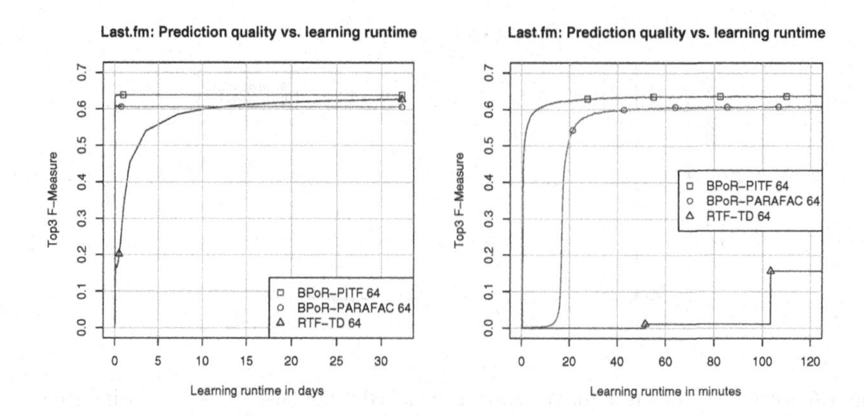

Fig. 7.7 F-Measure on top-3 list after training a model for x days/ hours. Learning a high quality TD model (RTF-TD) on a larger dataset like Last.fm takes several days. The PITF and PARAFAC models give good prediction quality already after 20 and 40 minutes respectively.

for Last.fm. With smaller factorization, the number of tags has a larger influence on the runtime – e.g. for $k = 16$ it is 0.3 ms vs. 1.1 ms. For the very large factorization of $k = 128$ and the very small dataset of BibSonomy, the prediction runtime of TD is worse than that of Folkrank (82.1 ms vs 19.1 ms). The reason is that the runtime of FolkRank depends on the size of the dataset – i.e. the observations S – and on the very small BibSonomy dataset that leads to a reasonable runtime but already for the larger Last.fm dataset the runtime of FolkRank is not feasible any more for real-time predictions. These results match to the theoretical prediction complexity which is $O(\text{iter}(|S|+|U|+|I|+|T|)+|T|N)$ for Folkrank (Jäschke et al, 2008) but only $O(|T|\cdot k_T + k_U \cdot k_I \cdot k_T)$ for TD.

Another major advantage of factorization models is that the tradeoff between quality and speed can be chosen by controlling the number of dimensions. That means depending on the application one can chose if runtime is more important than quality and thus reduce the number of dimensions. With FolkRank this tradeoff cannot be controlled.

The drawback of factorization models in comparison to FolkRank is that they need a training phase. But training is usually done offline and for online updating a factorization model there are very promising results for the related model class of regularized matrix factorization (Rendle and Schmidt-Thieme, 2008). Next, we investigate the training speed of various factorization models.

Learning runtime

The comparison of the convergence of BPoR-PITF to BPoR-PARAFAC and RTF-TD on the Last.fm dataset can be found in figure 7.7. Here you can see how the prediction quality improves after training a model (k=64) for a given time span. The left chart shows the quality over a span of 30 days. RTF-TD needs about 12 days to achieve a prediction quality as good as BPoR-PARAFAC. Even after 30 days of training, the quality of RTF-TD is still worse than BPoR-PITF.

In contrast to this, BPoR-PITF and BPoR-PARAFAC converge much faster. The right chart shows the quality over the first two hours. BPoR-PITF and BPoR-PARAFAC achieve convergence already after 20 and 40 minutes respectively. As each iteration of RTF-TD takes more than 50 minutes, the progress is very slow. When comparing BPoR-PITF and BPoR-PARAFAC among each other, one can see, that BPoR-PITF converges faster. It is interesting to see that in the beginning BPoR-PARAFAC seems to need several updates (18 minutes) before the quality improves reasonably. One explanation could be that BPoR-PARAFAC is searching the structure among the three-way interactions whereas in BPoR-PITF this is already given by the two pairwise interactions.

The worse empirical runtime results of RTF-TD in comparison to BPoR-PARAFAC and BPoR-PITF match to the theoretical runtime complexity analysis of the model equations (see chapter 5). Furthermore, learning for both BPoR-PARAFAC and BPoR-PITF can be easily parallelized because quadruples of two draws usually share no parameters – in contrast to this, all entries in RTF-TD share the core tensor which makes it more difficult to parallelize RTF-TD.

7.7 Conclusion

In this work, we have applied the context-aware ranking framework to the task of tag recommendation. Therefore, we have derived *Bayesian Post-aware Ranking* from BCR which leads to the optimization criterion BPoR-Opt and the learning algorithm BPoR-Learn. As models we have applied the PITF and PARAFAC models that have linear complexity. Furthermore, we have introduced RTF which optimizes the TD model for post-wise AUC and have shown how to speedup learning for RTF.

We have compared our factorization approaches to the state-of-the-art personalized tag recommenders FolkRank, PageRank and HOSVD as well as to the upper bound for any non-personalized tag recommender. Our results indicate that BPoR-PITF and RTF-TD have the best prediction quality outperforming all other approaches. Furthermore, our factorization methods have a much better prediction runtime which makes them applicable for real-world scenarios. Finally, we have empirically shown that our model PITF largely outperforms RTF-TD in learning runtime and achieves better prediction quality on datasets of large scale. The empirical comparison was done on lab experiments and on the 'ECML/ PKDD Discovery Challenge 2009', that PITF has won.

References

Heymann, P., Ramage, D., Garcia-Molina, H.: Social tag prediction. In: SIGIR 2008: Proceedings of the 31st annual International ACM SIGIR Conference on Research and Development in Information Retrieval, pp. 531–538. ACM, New York (2008)

Hotho, A., Jäschke, R., Schmitz, C., Stumme, G.: Information retrieval in folksonomies: Search and ranking. In: Sure, Y., Domingue, J. (eds.) ESWC 2006. LNCS, vol. 4011, pp. 411–426. Springer, Heidelberg (2006)

Jäschke, R., Marinho, L., Hotho, A., Schmidt-Thieme, L., Stumme, G.: Tag recommendations in folksonomies. In: Proceedings of the 11th European Conference on Principles and Practice of Knowledge Discovery in Databases (PKDD), Warsaw, Poland (2007)

Jäschke, R., Marinho, L., Hotho, A., Schmidt-Thieme, L., Stumme, G.: Tag recommendations in social bookmarking systems. AICOM (2008)

Ju, S., Hwang, K.B.: A weighting scheme for tag recommendation in social bookmarking systems. In: Proceedings of the ECML-PKDD Discovery Challenge Workshop (2009)

Lathauwer, L.D., Moor, B.D., Vandewalle, J.: A multilinear singular value decomposition. SIAM J. Matrix Anal. Appl. 21(4), 1253–1278 (2000)

Lathauwer, L.D., Moor, B.D., Vandewalle, J.: On the best rank-1 and rank-(r1,r2,.,rn) approximation of higher-order tensors. SIAM J. Matrix Anal. Appl. 21(4), 1324–1342 (2000)

Lipczak, M., Hu, Y., Kollet, Y., Milios, E.: Tag sources for recommendation in collaborative tagging systems. In: Proceedings of the ECML-PKDD Discovery Challenge Workshop (2009)

Marinho, L.B., Preisach, C., Schmidt-Thieme, L.: Relational classification for personalized tag recommendation. In: Proceedings of the ECML-PKDD Discovery Challenge Workshop (2009)

Rendle, S., Schmidt-Thieme, L.: Online-updating regularized kernel matrix factorization models for large-scale recommender systems. In: RecSys 2008: Proceedings of the 2008 ACM Conference on Recommender Systems, pp. 251–258. ACM, New York (2008)

Rendle, S., Schmidt-Thieme, L.: Factor models for tag recommendation in bibsonomy. In: Proceedings of the ECML-PKDD Discovery Challenge Workshop (2009)

Rendle, S., Marinho, L.B., Nanopoulos, A., Schmidt-Thieme, L.: Learning optimal ranking with tensor factorization for tag recommendation. In: KDD 2009: Proceeding of the 15th ACM SIGKDD International Conference on Knowledge Discovery and Data Mining. ACM, New York (2009)

Rennie, J.D.M., Srebro, N.: Fast maximum margin matrix factorization for collaborative prediction. In: ICML 2005: Proceedings of the 22nd International Conference on Machine learning, pp. 713–719. ACM, New York (2005)

Song, Y., Zhang, L., Giles, C.L.: A sparse gaussian processes classification framework for fast tag suggestions. In: CIKM 2008: Proceeding of the 17th ACM Conference on Information and knowledge Management, pp. 93–102. ACM, New York (2008)

Song, Y., Zhuang, Z., Li, H., Zhao, Q., Li, J., Lee, W.C., Giles, C.L.: Real-time automatic tag recommendation. In: SIGIR 2008: Proceedings of the 31st Annual International ACM SIGIR Conference on Research and Development in Information Retrieval, pp. 515–522. ACM, New York (2008)

Symeonidis, P., Nanopoulos, A., Manolopoulos, Y.: Tag recommendations based on tensor dimensionality reduction. In: RecSys 2008: Proceedings of the 2008 ACM Conference on Recommender Systems, pp. 43–50. ACM, New York (2008)

Tucker, L.: Some mathematical notes on three-mode factor analysis. Psychometrika 31, 279–311 (1966)

Wetzker, R., Said1, A., Zimmermann, C.: Understanding the user: Personomy translation for tag recommendation. In: Proceedings of the ECML-PKDD Discovery Challenge Workshop (2009)

Wetzker, R., Zimmermann, C., Bauckhage, C., Albayrak, S.: I tag, you tag: translating tags for advanced user models. In: WSDM 2010: Proceedings of the third ACM International Conference on Web search and Data Mining, pp. 71–80. ACM, New York (2010)
Zhang, N., Zhang, Y., Tang, J.: A tag recommendation system based on graph. In: Proceedings of the ECML-PKDD Discovery Challenge Workshop (2009)

Chapter 8
Sequential-Set Recommendation

Our methods so far ignore time which is an important variable that can be monitored in almost any application. In this chapter, we extend item recommendation (see chapter 6) with time information. In general, time is a continuous variable with infinite support. Thus, our factorization models in chapter 5 cannot be applied directly as they assume a categorical domain. Also simple discretization of the domain would not work because (1) factorization models assume no a priori relationship between two variable instances (e.g. two close points in time) and (2) the model could not predict in the future as no observations for these variables are present. Thus, our approach is different: we reformulate the problem with sequences and use the independence assumptions of Markov chains within our model. That means for each user, we see his action of the past as a sequence – e.g. what products he has bought. Typically, several products are bought at the same day and thus, we have per user a sequences of sets (=baskets/ shopping carts). The Markov chain assumption is now that the next action (=shopping cart) of the user depends only on a few of his previous ones.

Markov chains (MC) and matrix factorization (MF) are two of the most popular approaches for item recommenders. As we have seen in chapter 6, MF methods learn the general taste of a user by factorizing the matrix over observed user-item preferences. On the other hand, MC methods model sequential behavior by learning a transition graph over items that is used to predict the next action based on the recent actions of a user. Both MF and MC have their advantages: MF uses all data to learn the general taste of the user whereas MC can capture sequence effects in time by using a non-personalized transition matrix, i.e. the transition matrix is learned over all data of all users. In this chapter, we present a method bringing both approaches together. Our method is based on personalized transition graphs over underlying Markov chains. That means for each user an own transition matrix is learned – thus in total the method uses a transition cube. As the observations for estimating the transitions are usually very limited, our method factorizes the transition cube with the pairwise interaction model. We show that our factorized personalized MC (FPMC) model subsumes both a common Markov chain and the normal matrix factorization model. For learning the model parameters, we introduce an adaption of

S. Rendle: Context-Aware Ranking with Factorization Models, SCI 330, pp. 113–133.
springerlink.com

the Bayesian Context-aware Ranking (BCR) for sequential basket data. Empirically, we show that our FPMC model outperforms both the common matrix factorization and the unpersonalized MC model learned with and without factorization.

In total the contributions are as follows:

- We introduce personalized Markov chains relying on personalized transition matrices. This allows to capture both sequential effects and long term user-taste. We show that this is a generalization of both standard MC and MF models.
- To deal with the sparsity for the estimation of transition probabilities, we introduce a factorization model that can be applied both to personalized and normal transition matrices. This factorization approach results in less parameters and due to generalization to a better quality than full parametrized models.
- We empirically show that our model outperforms other state-of-the-art methods on sequential data.

8.1 Related Work

Markov chains for recommender systems have been studied by several researchers. Zimdars et al (2001) describe a sequential recommender based on Markov chains. They investigate how to extract sequential patterns to learn the next state with a standard predictor – e.g. a decision tree. Mobasher et al (2002) use pattern mining methods to discover sequential patterns which are used for generating recommendations. Shani et al (2005) introduce a recommender based on Markov decision processes (MDP) and also a MC based recommender. To enhance the maximum likelihood estimates (MLE) of the MC transition graphs, they describe several heuristic approaches like clustering and skipping. Instead of improving the MLE estimates with heuristics, we use a factorization model that is learned for optimal ranking instead of transition MLE. In total, the main difference of our work to all the previous approaches is the use of personalized transition graphs which bring together the benefits of sequential, i.e. time-aware, MC with time-invariant user taste. Furthermore factorizing transition probabilities and optimizing the parameters for ranking is new.

On the other hand, most of the recommender systems do not take sequential patterns into account and recommend based on the whole user history. Besides a very large literature on rating prediction (i.e. regression) emerging from the Netflix contest (e.g. Koren (2008, 2009)), item recommendation from implicit feedback has started to get more into the focus. Three recent methods (Hu et al, 2008; Pan and Scholz, 2009; Rendle et al, 2009) are based on the matrix factorization model that factorizes the matrix of user-item correlations (see chapter 6 for more details). In this work, we will bring the advantages of these MF models together with MC models.

8.2 Item Recommendation from Sequential Set Data

Item recommendation is the task of suggesting a personalized list of items (e.g. products, songs) for a specific user. This can be seen as creating a personalized

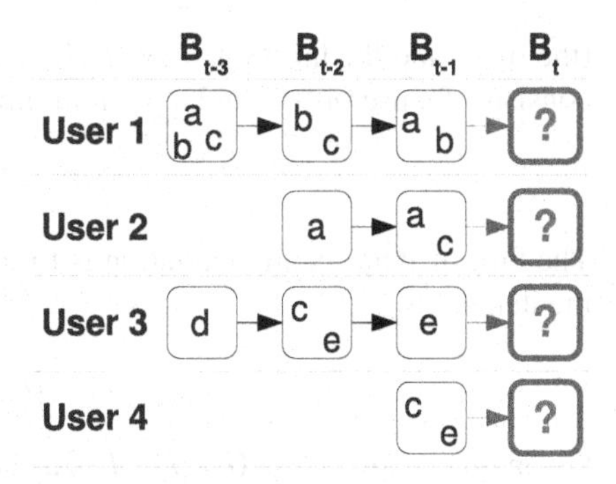

Fig. 8.1 Sequential basket data with four users and five items $\{a,b,c,d,e\}$. The task is to recommend items at time t given a basket history $B_{t-1}, B_{t-2}, \ldots$.

ranking on the items. Usually, recommender systems rely on statistical models that use the event history (e.g. purchases, listening) of users on items to generate recommendations. Time and thus sequential behavior is an important additional information that is tracked in almost any real-world application. Secondly, we consider the problem setting with set data – e.g. in online-shopping usually a basket of products is bought at the same time. In total, our setting is item recommendation from sequential set data. An example of such data can be found in figure 8.1.

8.2.1 Sequential vs. General Recommender

The most common approach to generate recommendations is to discard any sequential information and to learn what items a user is interested in in general. On the other hand, recommendations of sequential methods (mostly relying on Markov chains) are based only on the user's last events by learning in general what someone buys next when he bought a certain item in the recent past. Both methods have their strengths and disadvantages. Imagine a user that in general buys movies like 'Star Trek' and 'Star Wars'. In contrast to his usual buying behavior, he recently has purchased 'Titanic' and 'Dirty Dancing' to watch with his girlfriend. After that a MC based recommender of length 2 would only recommend movies like 'Notting Hill' and other romantic movies. In contrast, a global personalized recommender would correctly factor in the general taste of the user and recommend also movies like 'Back To the Future', 'Alien' or other science fiction movies. But there are also examples where sequential recommenders have advantages: E.g. good recommendations for a user that has recently bought a digital camera are accessories that other users have bought after buying that camera – this is exactly what a Markov chain model does. Global personalized recommender would not adapt directly to the recent purchase (the digital camera) but would recommend items this user likes in general.

8.2.2 Formalization

Item recommendation with time information is a three-mode problem ($m = 3$). We will refer to the first domain as $U = \{u_1, u_2, \ldots\}$ (the users), the second one es T

(the time) and the third one $I = \{i_1, i_2, \ldots\}$ (the items/ products). Here, the context consists of a user $u \in U$ and a point in time $t \in T$:

$$\mathscr{C} = U \times T \tag{8.1}$$

The task of item recommendation is to find a user and time specific ranking $\succ$ of the items:

$$\succ \subset U \times T \times I^2 \tag{8.2}$$

Given observations $S \subseteq U \times T \times I$, we can formulate the concept of *baskets* $B \subseteq I$. Now, the training data S can be seen as a set $\mathscr{B}$ of sequences of baskets:

$$\mathscr{B} := \{\mathscr{B}^{u_1}, \ldots \mathscr{B}^{u_{|U|}}\} \tag{8.3}$$

where $\mathscr{B}^u$ is the basket sequence of a user:

$$\mathscr{B}^u = (B^u_{t_1}, B^u_{t_2}, \ldots), \quad t_a < t_b \Leftrightarrow a < b \tag{8.4}$$

And each basket is defined by S:

$$B^u_t = \{i : (u,t,i) \in S\} \tag{8.5}$$

In general, the time domain is continuous and infinite, i.e. $T = \mathbb{R}$. In this work, we are not interested in predicting rankings for a specific quantitative point in time (i.e. 1st January 2010), but only in qualitative ones – e.g. the first, second, etc. basket of a user. For each user, we are interested in ranking his next basket – independently of when 'next' happens exactly. This means, we can see the domain of T as the natural numbers ($T = \mathbb{N}$) where two points $t_1, t_2 \in T$ only have a semantic within a user – e.g. the third basket (B^u_3) of user u was bought after the second basket (B^u_2) of user u, but we cannot say that it (B^u_3) was bought after the second basket ($B^{u'}_2$) of another user u'.

Usually, time is observed as reals ($T = \mathbb{R}$). But with a simple mapping, we can transform this to our notation. Let ϕ be the user-specific time mapping $\phi : U \times \mathbb{R} \to \mathbb{N}$ from quantitative time to user-specific qualitative time:

$$\phi(u,t) := |\{t' \in (-\infty, t) : (u,t',i) \in S\}| \tag{8.6}$$

Thus, in the following we will only work with the sequential view of the problem and the task is to predict the next basket for a user at time t, where $t \in \mathbb{N}$.

8.2.3 Modelling and Estimation

Like usual, we want to model and estimate a function $y : U \times T \times I \to \mathbb{R}$ that can be used to express the context-aware ranking $\succ$ according to eq. (3.34). Even though with the sequential view the time T is finite, modelling y directly with the tensor factorization models (see chapter 5) makes no sense because the time T would be handled wrong:

1. Factorization models assume that there is no a priori relation between two different variable instances. E.g. a factorization model would not know that the baskets B_4^u and B_5^u are related.
2. Even worse, factorization models assume that two identical variable instances have identical factors. But by seeing time as a sequence *per* user, the t-th basket of user u_1 and t-th basket of u_2 have a priori nothing to do with each other.

Instead, we model y by a Markov chain: the estimated basket at time t depends on the basket of time $t-1$.

8.3 Factorizing Personalized Markov Chains (FPMC)

First, we introduce MC for sequential set data and extend this to personalized MCs. We discuss the weakness of Maximum Likelihood Estimates for the transition cubes. To solve this, we introduce factorized transition cubes where information among transitions is propagated. We conclude this section by combining both ideas into FPMCs.

8.3.1 Personalized Markov Chains for Sets

Before introducing personalized MCs, we first describe how to model the unpersonalized MC for sets with a reasonable state space. Then we show how to estimate the parameters for this unpersonalized MC with the maximum likelihood estimator (MLE). Afterwards, the extension of both the model and the estimation to personalized MCs is simple. Finally, we will show the limitations of full parametrized transition graphs (i.e. one parameter per transition) and the MLE method for personalized Markov chains.

8.3.1.1 Markov Chains for Sets

In general, a Markov chain of order q is defined as:

$$p(Z_t = z_t | Z_{t-1} = z_{t-1}, \ldots, Z_{t-q} = z_{t-q}) \tag{8.7}$$

Where $Z_t, \ldots, Z_{t-q}$ are random variables and z_{t-j} their realizations. In a recommender application without sets, the random variables are defined over I – i.e. realizations are single items $i \in I$. But in our case, the variables are defined over $\mathscr{P}(I)$ as the realizations are whole baskets B and thus the size of the state space is $2^{|I|}$. Obviously, defining a long chain over the whole state space is not feasible for sets. To handle this huge state space, we make two simplifications: (1) we use chains of length $q = 1$ and (2) the transition probabilities are simplified.

A Markov chain of order $q = 1$ for the basket problem is:

$$p(B_t|B_{t-1}) \tag{8.8}$$

In recommender scenarios without sets, usually longer chains (e.g. $q = 3$) are preferable (Shani et al, 2005) because a history with size $q = 1$ contains only one item. In our case with sets, even a chain with length $q = 1$ is reasonable because it relies already on many items (all items of the basket) – e.g. in the application of our evaluation there are about 10 items on average (see table 8.1).

Markov chains of length $q = 1$ are described by their stochastic transition matrix A over the state space. In our case the state space over sets is $\mathscr{P}(I)$ and thus the dimensionality of the transition matrix would be $2^{|I|} \times 2^{|I|}$. Thus, instead of modeling transition over baskets, we model transitions over $|I|$ binary variables that describe a set/ basket:

$$a_{l,i} := p(i \in B_t | l \in B_{t-1}) \tag{8.9}$$

Using this representation has the following implications:

- The state space is now I and thus the size of the transition matrix A is $|I|^2$ – by factorization, we will later further reduce the number of parameters needed to represent this space from $|I|^2$ to $2k\,|I|$ where k is the number of latent dimensions used in the factorization model.
- The elements of the state space are $i \in B$ which is a binary variable, thus $p(i \in B_t | l \in B_{t-1}) + p(i \notin B_t | l \in B_{t-1}) = 1$. This means, that the transition matrix A is no longer stochastic, because $\sum_{i \in I} a_{l,i} = 1$ does not hold.

For item recommendation (i.e. for estimating the ranking function y), we are interested in the probability of purchasing an item given the last basket of a user. This can be defined as the mean over all transitions from the purchases of the last basket to this item:

$$p(i \in B_t | B_{t-1}) := \frac{1}{|B_{t-1}|} \sum_{l \in B_{t-1}} p(i \in B_t | l \in B_{t-1}) \tag{8.10}$$

And the full Markov chain over baskets can be expressed by:

$$p(B_t|B_{t-1}) \propto \prod_{i \in I} p(i \in B_t | B_{t-1}) \tag{8.11}$$

Note that in general we are not interested in the probabilities of the full Markov chain (eq. (8.11)), but rather in the probabilities of single items given the last basket (eq. (8.10)) because this can be used to recommend the most probable items.

8.3.1.2 Estimation of Transition Probabilities

To make predictions using the Markov chain in eq. (8.10), the transition probabilities $a_{l,i}$ have to be estimated. The maximum likelihood estimator for $a_{l,i}$ given the data $\mathscr{B}$ is:

$$\hat{a}_{l,i} = \hat{p}(i \in B_t | l \in B_{t-1}) = \frac{\hat{p}(i \in B_t \wedge l \in B_{t-1})}{\hat{p}(l \in B_{t-1})}$$
$$= \frac{|\{(B_t, B_{t-1}) : i \in B_t \wedge l \in B_{t-1}\}|}{|\{(B_t, B_{t-1}) : l \in B_{t-1}\}|} \tag{8.12}$$

An example for such a non-personalized MLE can be seen in figure 8.2. Here, the buying history for the four users of figure 8.1 are translated into transitions A of eq. (8.10). The transition matrix can then be applied to predict which items should be recommended given the last basket. E.g. for user 4, the probabilities would be:

$$p(a \in B_t | \{c,e\}) = 0.5(0.3 + 0.0) = 0.15$$
$$p(b \in B_t | \{c,e\}) = 0.5(0.7 + 0.0) = 0.35$$
$$p(c \in B_t | \{c,e\}) = 0.5(0.3 + 0.0) = 0.15$$
$$p(d \in B_t | \{c,e\}) = 0.5(0.0 + 0.0) = 0.00$$
$$p(e \in B_t | \{c,e\}) = 0.5(0.3 + 1.0) = 0.65$$

As the user has already bought item c and e, the best recommendation of unknown items would be b and then a. Looking only at the items this and similar users have bought in the past, one would expect, that item d might be a better recommendation.

8.3.1.3 Personalized Markov Chains for Sets

Until now, the MC has been defined unpersonalized, i.e. independently of the user. Next, we extend this to a personalized MC per user:

$$p(B_t^u | B_{t-1}^u) \tag{8.13}$$

Again, we represent each MC by the transitions over items, but now user-specific:

$$a_{u,l,i} := p(i \in B_t^u | l \in B_{t-1}^u) \tag{8.14}$$

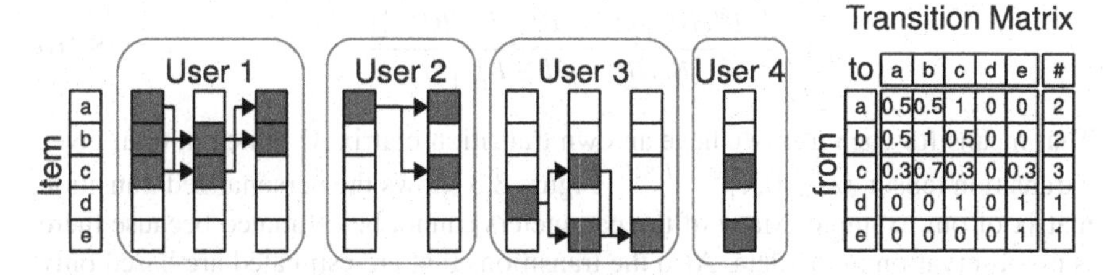

to	a	b	c	d	e	#
a	0.5	0.5	1	0	0	2
b	0.5	1	0.5	0	0	2
c	0.3	0.7	0.3	0	0.3	3
d	0	0	1	0	1	1
e	0	0	0	0	1	1

Fig. 8.2 Non-personalized Markov chain: The transition matrix contains the MLE estimates for the probability $p(i \in B_t | l \in B_{t-1})$ using the data of figure 8.1. The column # states how many observations were used to estimate this transition. In this example, the users 1 and 2 as well as 3 and 4 share a similar taste for items a, c and items c, e respectively. Thus, one would expect to find d before b in the recommendation list for user 4, but the MC would recommend b as best unknown item.

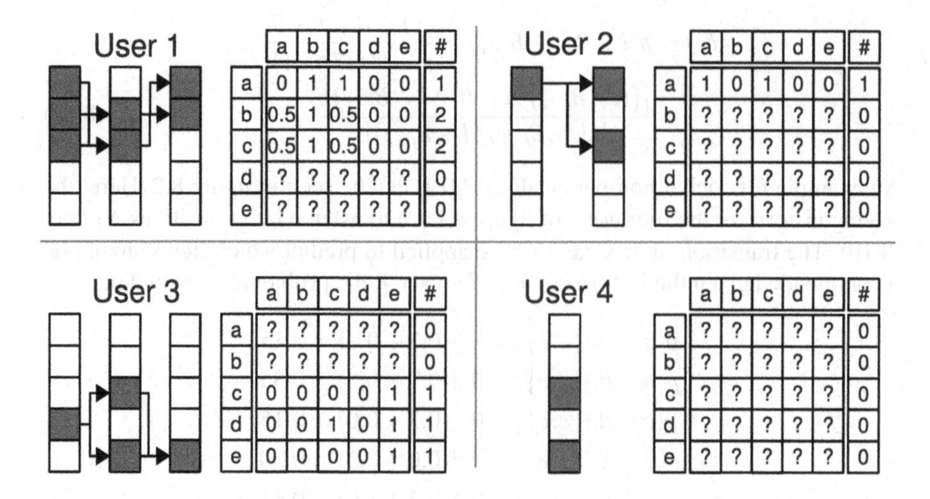

Fig. 8.3 Personalized Markov chains: For each user an individual transition matrix is given. The transition matrices contain the MLE estimates for the probability $p(i \in B_t^u | l \in B_{t-1}^u)$. Entries with ? are missing values as there is no data to estimate the probabilities. Obviously, estimating the personalized transition matrices directly results in very poor transitions as each estimate is not reliable. This problem will be solved later by factorizing the transitions.

And thus also the prediction depends only on the user's transitions:

$$p(i \in B_t^u | B_{t-1}^u) := \frac{1}{|B_{t-1}^u|} \sum_{l \in B_{t-1}^u} p(i \in B_t^u | l \in B_{t-1}^u) \tag{8.15}$$

Also MLE can be applied analogously but now the transitions for user u are only estimated from his history $\mathscr{B}^u$ – that means u is not a free variable anymore:

$$\begin{aligned} \hat{a}_{u,l,i} = \hat{p}(i \in B_t^u | l \in B_{t-1}^u) &= \frac{\hat{p}(i \in B_t^u \wedge l \in B_{t-1}^u)}{\hat{p}(l \in B_{t-1}^u)} \\ &= \frac{|\{(B_t^u, B_{t-1}^u) : i \in B_t^u \wedge l \in B_{t-1}^u\}|}{|\{(B_t^u, B_{t-1}^u) : l \in B_{t-1}^u\}|} \end{aligned} \tag{8.16}$$

That means for each user we have an own transition matrix A^u which in total gives a transition tensor $\mathscr{A} \in [0,1]^{|U| \times |I| \times |I|}$. Figure 8.3 shows the personalized transition matrix of our example. Many of the parameters cannot be estimated because there is no observation in the data. Also the transitions that are estimated are based only on a small number of observations that means they are unreliable. At first glance, using personalized MCs seems to be unreasonable. We will discuss next what are the reasons for the poor estimations and show how to fix it.

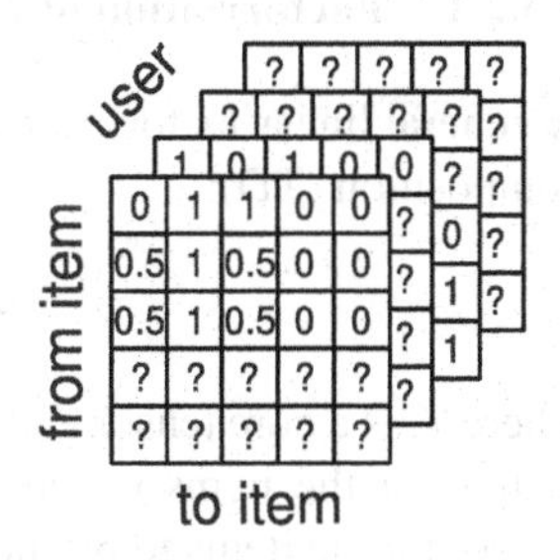

Fig. 8.4 Personalized transition cube: Stacking all transition matrices of the individual users leads to a transition cube. Instead of a fully parametrized cube which is very sparse, a factored cube is used to generate better transition estimates.

8.3.1.4 Limitations of MLE and Full Parametrization

The problem of unreliable transition probabilities both for unpersonalized and even more for personalized MCs lies in the fact that they work with a fully parametrized transition graph (e.g. matrix and tensor respectively) and the way of parameter estimation. Full parametrization means we have $|I|^2$ and $|U|\,|I|^2$ respectively independent parameters for describing the transitions. Note that the MLE estimates each transition parameter $a_{l,i}$ independently from the others, i.e. none of the cooccurrences (l,i) will contribute to another transition probability estimator (l,j) but only to $p(i \in B_t | l \in B_{t-1})$. This is even worse for personalized MCs as a triple (u,l,i) does not contribute to the estimate of (u',l,i). In addition, the important properties of MLE (e.g. Gaussian distribution, unbiased estimator, minimal variance under all unbiased estimators) only exist in asymptotic theory. In cases of less data they suffer from underfitting. Since in our scenario the data is extremely sparse, Maximum Likelihood Estimators easily fail.

To get more reliable estimates for the transitions, we factorize the transition cube which breaks the independence of the parameters and the estimation. This way, each transition is influenced by similar users, similar items and similar transitions because information propagates through this model. In our evaluation, we show that this way (1) better transition graphs than MLE can be generated for the non-personalized setting and (2) that personalized MCs outperform both non-personalized factorized MC and non-personalized full parameterized MLE MCs.

8.3.2 Factorizing Transition Graphs

In the following, we will derive a factorization model for the transition cube $\mathscr{A}$. That means we model the unobserved transition tensor $\mathscr{A}$ by a low rank approximation $\hat{\mathscr{A}}$. The advantage of this approach over a full parametrization is that it can handle sparsity and generalizes to unobserved data because information propagates through the model – i.e. parameters influence each other.

8.3.2.1 Factorization of the Transition Cube

A general linear factorization model for estimating the tensor $\mathscr{A}$ is the Tucker Decomposition (TD):

$$\hat{\mathscr{A}} := \mathscr{C} \times_U V^U \times_L V^L \times_I V^I \tag{8.17}$$

where $\mathscr{C}$ is a core tensor and V^U is the feature matrix for the users, V^L is the feature matrix for the items in the last transition (outgoing nodes) and V^I is the feature matrix for the items to predict (ingoing nodes). They have the following structure:

$$\mathscr{C} \in \mathbb{R}^{k_U,k_L,k_I}, \quad V^U \in \mathbb{R}^{|U|\times k_U}, V^L \in \mathbb{R}^{|I|\times k_L}, \quad V^I \in \mathbb{R}^{|I|\times k_I} \tag{8.18}$$

with the factorization dimensions k_U, k_L and k_I. In chapter 5, we have shown that the Tucker decomposition subsumes other factorization models like parallel factor analysis (PARAFAC) and the pairwise interaction model (PITF).

As the observed transitions for $\mathscr{A}$ are very sparse, we use the PITF model:

$$\hat{a}_{u,l,i} := \langle \mathbf{v}_u^{U,I}, \mathbf{v}_i^{I,U} \rangle + \langle \mathbf{v}_i^{I,L}, \mathbf{v}_l^{L,I} \rangle + \langle \mathbf{v}_u^{U,L}, \mathbf{v}_l^{L,U} \rangle \tag{8.19}$$

or equivalently:

$$\hat{a}_{u,l,i} := \sum_{f=1}^{k_{U,I}} v_{u,f}^{U,I} v_{i,f}^{I,U} + \sum_{f=1}^{k_{I,L}} v_{i,f}^{I,L} v_{l,f}^{L,I} + \sum_{f=1}^{k_{U,L}} v_{u,f}^{U,L} v_{l,f}^{L,U} \tag{8.20}$$

This model directly models the pairwise interaction between all three modes of the tensor, i.e. between U and I, U and J as well as J and I. In total for each mode (i.e. user U, item I, item J), we have two factorization matrizes:

1. For the interaction between U and I: $V^{U,I} \in \mathbb{R}^{|U|\times k_{U,I}}$ modelling the user features and $V^{I,U} \in \mathbb{R}^{|I|\times k_{U,I}}$ for the last item i.
2. For the interaction between I and L: $V^{I,L} \in \mathbb{R}^{|I|\times k_{I,L}}$ for the next item i and $V^{L,I} \in \mathbb{R}^{|I|\times k_{I,L}}$ for the last item l.
3. For the interaction between U and L: $V^{U,L} \in \mathbb{R}^{|U|\times k_{U,L}}$ for the user features and $V^{L,U} \in \mathbb{R}^{|I|\times k_{U,L}}$ for the features of the last item l.

An advantage of this model over TD is that the prediction and learning complexity is much lower than for TD (see table 5.1). Furthermore even though TD and PARAFAC subsume the pairwise interaction model, with standard regularization estimation procedures have problems identifying such a model (see chapter 7). Finally, by using PITF we will later show that the factorized personalized markov chain subsumes the common matrix factorization model for item recommendation. This would also hold when using TD or PARAFAC (because TD and PARAFAC subsume PITF), but the analogies will be more obvious with PITF.

In section 8.4 we describe how to optimize the model parameters (factorization matrices) for item recommendation.

8.3.2.2 Factorization of the Transition Matrix

The proposed model for factorizing transition cubes can also be applied to estimate a transition matrix A (see formula (8.9)) for cases where no personalization of the transition graph is desired. By skipping the user-interactions in equation (8.19), a factorization model for normal transition graphs is obtained:

$$\hat{a}_{l,i} := \langle \mathbf{v}_i^{I,L}, \mathbf{v}_l^{L,I} \rangle \tag{8.21}$$

Also the parameter estimation method in section 8.4 can be used for optimizing the factorization matrices.

8.3.3 *Summary of FPMC*

Bringing together the personalized set MC (eq. (8.15)) with the factorized transition cube (eq. (8.19)) results in the factorized personalized Markov chain (FPMC):

$$p(i \in B_t^u | B_{t-1}^u) = \frac{1}{|B_{t-1}^u|} \sum_{l \in B_{t-1}^u} p(i \in B_t^u | l \in B_{t-1}^u) \tag{8.22}$$

We model $p(i \in B_t^u | l \in B_{t-1}^u)$ with the factorization cube $\hat{\mathscr{A}}$:

$$\begin{aligned} \hat{p}(i \in B_t^u | B_{t-1}^u) &= \frac{1}{|B_{t-1}^u|} \sum_{l \in B_{t-1}^u} \hat{a}_{u,l,i} \\ &= \frac{1}{|B_{t-1}^u|} \sum_{l \in B_{t-1}^u} \left(\langle \mathbf{v}_u^{U,I}, \mathbf{v}_i^{I,U} \rangle + \langle \mathbf{v}_i^{I,L}, \mathbf{v}_l^{L,I} \rangle + \langle \mathbf{v}_u^{U,L}, \mathbf{v}_l^{L,U} \rangle \right) \end{aligned} \tag{8.23}$$

And as the factorization (U,I) is independent of l, we can remove it from the sum:

$$\hat{p}(i \in B_t^u | B_{t-1}^u) = \langle \mathbf{v}_u^{U,I}, \mathbf{v}_i^{I,U} \rangle + \frac{1}{|B_{t-1}^u|} \sum_{l \in B_{t-1}^u} \left(\langle \mathbf{v}_i^{I,L}, \mathbf{v}_l^{L,I} \rangle + \langle \mathbf{v}_u^{U,L}, \mathbf{v}_l^{L,U} \rangle \right) \tag{8.24}$$

In the next section, we apply this model to the task of item recommendation. We will show that in this case, the model can be simplified even more because the interaction between U and L vanishes.

Besides better generalization of factorization models compared to a fully parametrized transition cube, a further advantage is that less parameters are needed. Instead of $|U| \cdot |I|^2$ parameters in a full parametrized cube or $|I|^2$ in a full parametrized matrix, the factorization model only needs $2 \cdot k_{I,L} \cdot |I|$ parameters for the non-personalized model and $2 \cdot k_{I,L} \cdot |I| + k_{U,I} \cdot (|U| + |I|)$ parameters for the personalized model. This is especially important for applications with a high number of items where a full parametrization with $|I|^2$ parameters might not be feasible.

8.4 Item Recommendation with FPMC

So far, a factorization model for personalized Markov chains has been introduced. In the following, we will apply this model to the task of item recommendation. That means, the model parameters should be optimized for ranking. First, we derive S-BPR which is a general optimization criterion for item recommendation from sequential set data. This optimization criterion is not limited to our FPMC model and can be applied also to other models like kNN or standard MF. Secondly, we apply S-BPR to FPMC and show how the model can be simplified in the case of item recommendation using S-BPR. Afterwards we present a stochastic gradient descent learning algorithm based on bootstrap sampling for optimizing the model parameters with S-BPR.

8.4.1 *Optimization Criterion S-BPR*

As described in section 8.2, the goal of item recommendation from sequential basket data is to derive a ranking $>_{u,t}$ over the items. To model the ranking, we assume there is an estimator $\hat{y} : U \times T \times I \to \mathbb{R}$ – e.g. the buying probability of the personalized Markov Chain – which is used to define the ranking:

$$i >_{u,t} j \;:\Leftrightarrow\; \hat{y}(u,t,i) > \hat{y}(u,t,j) \tag{8.25}$$

See section 3.4 for more details.

Next, we derive the sequential BPR (S-BPR) optimization criterion analogously to the general BCR approach (section 4.1). The probability of the model parameter Θ given the ranking $>_{u,t} \subset I \times I$ for user u at time t can be formalized as:

$$p(\Theta| >_{u,t}) \propto p(>_{u,t} |\Theta)\, p(\Theta) \tag{8.26}$$

In our case the model parameters are $\Theta = \{V^{U,I}, V^{I,U}, V^{L,I}, V^{I,L}, V^{U,L}, V^{L,U}\}$.

Assuming independence of baskets and users given Θ, this leads to the maximum a posteriori (MAP) estimator of the model parameters:

$$\underset{\Theta}{\operatorname{argmax}} \prod_{u \in U} \prod_{B_t \in \mathscr{B}^u} p(>_{u,t} |\Theta)\, p(\Theta) \tag{8.27}$$

Expanding $>_{u,t}$ for all item-pairs $(i, j) \in I^2$ and using the same assumptions as in section 4.1, the probability of $p(>_{u,t} |\Theta)$ can be rewritten as:

$$\prod_{u \in U} \prod_{B_t \in \mathscr{B}^u} \prod_{i \in B_t} \prod_{j \notin B_t} p(i >_{u,t} j|\Theta)^2 \tag{8.28}$$

Next we use the model definition of eq. (8.25) to express $p(i >_{u,t} j|\Theta)$:

$$p(i >_{u,t} j|\Theta) = p(\hat{y}(u,t,i) > \hat{y}(u,t,j)\,|\,\Theta) = p(\hat{y}(u,t,i) - \hat{y}(u,t,j) > 0\,|\,\Theta) \tag{8.29}$$

The model parameters Θ can be skipped as $\hat{y}$ contains them implicitly – i.e. $\hat{y} = \hat{y}(\Theta)$. And furthermore we define $p(z > 0) := \sigma(z)$ using the logistic sigmoid function σ:

$$p(i \succ_{u,t} j|\Theta) = \sigma(\hat{y}(u,t,i) - \hat{y}(u,t,j)) \tag{8.30}$$

Furthermore, we assume Gaussian priors on the model parameters: $\theta \sim \mathcal{N}(0, \frac{1}{2\lambda_\theta})$.

In total this leads to the MAP-estimator for sequential BPR:

$$\begin{aligned}
&\operatorname*{argmax}_{\Theta} \ln p(\succ_{u,t} |\Theta)\, p(\Theta) \\
&= \operatorname*{argmax}_{\Theta} \ln \prod_{u\in U} \prod_{B_t \in \mathscr{B}^u} \prod_{i\in B_t} \prod_{j \notin B_t} \sigma(\hat{y}(u,t,i) - \hat{y}(u,t,j))^2 p(\Theta) \\
&= \operatorname*{argmax}_{\Theta} \sum_{u\in U} \sum_{B_t \in \mathscr{B}^u} \sum_{i\in B_t} \sum_{j \notin B_t} \ln \sigma(\hat{y}(u,t,i) - \hat{y}(u,t,j)) - \sum_{\theta\in\Theta} \frac{1}{2} \lambda_\theta\, \theta^2 \\
&=: \operatorname*{argmax}_{\Theta} \text{SBPR-O}\textsc{pt}
\end{aligned} \tag{8.31}$$

This can also be written with our standard notation of D_S (see section 3.3.1):

$$\begin{aligned}
&\operatorname*{argmax}_{\Theta} \text{SBPR-O}\textsc{pt} \\
&= \operatorname*{argmax}_{\Theta} \sum_{(u,t,i,j)\in D_S} \ln \sigma(\hat{y}(u,t,i) - \hat{y}(u,t,j)) - \sum_{\theta\in\Theta} \frac{1}{2} \lambda_\theta\, \theta^2
\end{aligned} \tag{8.32}$$

8.4.2 Item Recommendation from Sequential Set Data with FPMC

For item recommendation with FPMC, we express $\hat{y}$ by the FPMC model and apply S-BPR. We will show that one of the pairwise effects of FPMC vanishes which leads to a more compact model.

First, we use FPMC to express $\hat{y}$:

$$\begin{aligned}
\hat{y}'(u,t,i) &:= \hat{p}(i \in B^u_t | B^u_{t-1}) \\
&= \langle \mathbf{v}^{U,I}_u, \mathbf{v}^{I,U}_i \rangle + \frac{1}{|B^u_{t-1}|} \sum_{l\in B^u_{t-1}} \left(\langle \mathbf{v}^{I,L}_i, \mathbf{v}^{L,I}_l \rangle + \langle \mathbf{v}^{U,L}_u, \mathbf{v}^{L,U}_l \rangle \right)
\end{aligned} \tag{8.33}$$

Lemma 8.1 (Invariance of (U,L) decomposition). *For ranking of items and optimization with S-BPR, the FPMC model is invariant to the (U,L) decomposition, i.e. $\hat{y}'$ is invariant to $\hat{y}$ with:*

$$\hat{y}(u,t,i) := \langle \mathbf{v}^{U,I}_u, \mathbf{v}^{I,U}_i \rangle + \frac{1}{|B^u_{t-1}|} \sum_{l\in B^u_{t-1}} \langle \mathbf{v}^{I,L}_i, \mathbf{v}^{L,I}_l \rangle \tag{8.34}$$

Proof. Let $\mathbf{c} = (u,t)$, then:

$$\hat{y}'(\mathbf{c},i) = \hat{y}(\mathbf{c},i) + \hat{z}(\mathbf{c}) \tag{8.35}$$

with:

$$\hat{z}(\mathbf{c}) = \frac{1}{|B^u_{t-1}|} \sum_{l \in B^u_{t-1}} \langle \mathbf{v}^{U,L}_u, \mathbf{v}^{L,U}_l \rangle \tag{8.36}$$

Now, we can apply lemma 5.2 because S-BPR is an instance of BCR.

The lemma shows that in our case the (U,L) interaction can be dropped. And thus for item recommendation with FPMC the simpler model $\hat{y}$ from eq. (8.34) should be used.

8.4.2.1 Expressiveness

Next, we will show the analogies of the simplified FPMC model to standard matrix factorization (MF) and a factorized Markov chain (FMC). First, we will recollect the definitions of MF and FMC. In our notation, the standard Matrix factorization model for item recommendation (see chapter 6) is:

$$\hat{y}^{\text{MF}}(u,t,i) := \langle \mathbf{v}^{U,I}_u, \mathbf{v}^{I,U}_i \rangle \tag{8.37}$$

where $\hat{y}$ is independent of the sequential behaviour, i.e. independent of t.

Factorizing an unpersonalized Markov chain using equation (8.10) and (8.21) leads to:

$$\hat{y}^{\text{FMC}}(u,t,i) := \frac{1}{|B_{t-1}|} \sum_{l \in B_{t-1}} \langle \mathbf{v}^{I,L}_i, \mathbf{v}^{L,I}_l \rangle \tag{8.38}$$

Thus FPMC (eq. (8.34)) is a linear combination of both models:

$$\hat{y}^{\text{FPMC}}(u,t,i) = \hat{y}^{\text{MF}}(u,t,i) + \hat{y}^{\text{FMC}}(u,t,i) \tag{8.39}$$

This means FPMC can express both models: By setting the factorization dimensionality of (U,I) to zero ($k_{U,I} = 0$) a pure FMC is obtained and analogously setting $k_{I,L} = 0$ leads to a pure MF model.

It is important to note, that even though the model equation for FPMC in the case of item recommendation can be expressed by a combination of a MF and a FMC model, it is different from a simple ensemble of a single MF with a single FMC model because in our case the model parameters are learned jointly. Thus the learned model parameters jointly represent the personalized Markov chain instead of just pure user-item interactions and a global MC. This gets more obvious in the general case of FPMC where the model equation cannot be expressed by a linear combination of MC and FMC. Examples are (1) optimizing for another objective criterion (e.g. least-square) where the (U,L) decomposition cannot be dropped because here the invariance to the objective (Lemma 8.1) does not hold like in S-BPR. And (2) using another factorization model for $\mathscr{A}$ in FPMC than pairwise interaction (e.g. PARAFAC or TD) also leads to a different model equation even for item recommendation with S-BPR.

8.4.3 Learning Algorithm

Next, we adapt the learning algorithm BCR-LEARN to S-BPR and apply it to FPMC (see algorithm 6). As FPMC subsumes MF and FMC, both of these models can also be optimized for S-BPR with the provided algorithm. In each iteration a quadruple (u,t,i,j) is drawn consisting of an item i in the basket B_t^u of user u at time t and an item j that is not in this basket. Then gradient descent on S-BPR using this quadruple is performed. The gradients of S-BPR with respect to a model parameter θ and a given (u,t,i,j) are:

Algorithm 6 S-BPR-LEARN-FPMC

Input: training data S, learning rate α, regularization parameter λ
Output: model parameters $V^{U,I}, V^{I,U}, V^{I,L}, V^{L,I}$

1: draw $V^{U,I}, V^{I,U}, V^{I,L}, V^{L,I}$ from $\mathscr{N}(0, \sigma^2)$
2: **repeat**
3: draw (u,t,i) uniformly from S
4: draw j uniformly from $(I \setminus B_t^u)$
5: $\delta \leftarrow (1 - \sigma(\hat{y}(u,t,i) - \hat{y}(u,t,j)))$
6: **for** $f \in \{1,\ldots,k_{U,I}\}$ **do**
7: $v_{u,f}^{U,I} \leftarrow v_{u,f}^{U,I} + \alpha \left(\delta \, (v_{i,f}^{I,U} - v_{j,f}^{I,U}) - \lambda \, v_{u,f}^{U,I} \right)$
8: $v_{i,f}^{I,U} \leftarrow v_{i,f}^{I,U} + \alpha \left(\delta \, v_{u,f}^{U,I} - \lambda \, v_{i,f}^{I,U} \right)$
9: $v_{j,f}^{I,U} \leftarrow v_{j,f}^{I,U} + \alpha \left(-\delta \, v_{u,f}^{U,I} - \lambda \, v_{j,f}^{I,U} \right)$
10: **end for**
11: $\eta \leftarrow \frac{1}{|B_{t-1}^u|} \sum_{l \in B_{t-1}^u} v_{l,f}^{L,I}$
12: **for** $f \in \{1,\ldots,k_{I,L}\}$ **do**
13: $v_{i,f}^{I,L} \leftarrow v_{i,f}^{I,L} + \alpha \left(\delta \, \eta - \lambda \, v_{i,f}^{I,L} \right)$
14: $v_{j,f}^{I,L} \leftarrow v_{j,f}^{I,L} + \alpha \left(-\delta \, \eta - \lambda \, v_{j,f}^{I,L} \right)$
15: **for** $l \in B_{t-1}^u$ **do**
16: $v_{l,f}^{L,I} \leftarrow v_{l,f}^{L,I} + \alpha \left(\delta \, \frac{v_{i,f}^{I,L} - v_{j,f}^{I,L}}{|B_{t-1}^u|} - \lambda \, v_{l,f}^{L,I} \right)$
17: **end for**
18: **end for**
19: **until** convergence
20: **return** $V^{U,I}, V^{I,U}, V^{I,L}, V^{L,I}$

$$\frac{\partial}{\partial \theta}\left(\ln \sigma(\hat{y}(u,t,i)-\hat{y}(u,t,j))-\frac{1}{2}\lambda_\theta\,\theta^2\right)$$
$$=(1-\sigma(\hat{y}(u,t,i)-\hat{y}(u,t,j)))\frac{\partial}{\partial \theta}(\hat{y}(u,t,i)-\hat{y}(u,t,j))-\lambda_\theta\,\theta \tag{8.40}$$

with

$$\frac{\partial}{\partial v^{U,I}_{u,f}}(\hat{y}(u,t,i)-\hat{y}(u,t,j)) = v^{I,U}_{i,f} - v^{I,U}_{j,f} \tag{8.41}$$
$$\frac{\partial}{\partial v^{I,U}_{i,f}}(\hat{y}(u,t,i)-\hat{y}(u,t,j)) = v^{U,I}_{u,f} \tag{8.42}$$
$$\frac{\partial}{\partial v^{I,U}_{j,f}}(\hat{y}(u,t,i)-\hat{y}(u,t,j)) = -v^{U,I}_{u,f} \tag{8.43}$$
$$\frac{\partial}{\partial v^{L,I}_{l,f}}(\hat{y}(u,t,i)-\hat{y}(u,t,j)) = \frac{1}{|B^u_{t-1}|}(v^{I,L}_{i,f} - v^{I,L}_{j,f}) \tag{8.44}$$
$$\frac{\partial}{\partial v^{I,L}_{i,f}}(\hat{y}(u,t,i)-\hat{y}(u,t,j)) = \frac{1}{|B^u_{t-1}|}\sum_{l\in B^u_{t-1}} v^{L,I}_{l,f} \tag{8.45}$$
$$\frac{\partial}{\partial v^{I,L}_{j,f}}(\hat{y}(u,t,i)-\hat{y}(u,t,j)) = -\frac{1}{|B^u_{t-1}|}\sum_{l\in B^u_{t-1}} v^{L,I}_{l,f} \tag{8.46}$$

The complexity of the algorithm is $O(\#\text{it}\,(k_{U,I}+k_{I,L}\,\overline{|B|}))$ where $\overline{|B|}$ is the average basket size in $\mathscr{B}$ and #it is the number of iterations.

8.5 Evaluation

We empirically compare the recommender quality of our proposed factorized MC methods (factorized personalized Markov chain (FPMC) and factorized Markov chain (FMC)) to non-factorized Markov chain ('MC dense'), matrix factorization (MF) and the most-popular baseline – i.e. ranking all items by how often they have been bought in the past. Note that this comparison includes the strong baseline method BPR-MF (see chapter 6). As MF ($k_{I,L}=0$) and FMC ($k_{U,I}=0$) are a special case of FPMC, we use the FPMC learning algorithm for all three methods.

8.5.1 Dataset

The evaluation is done on anonymized purchase data of an online drug store[1]. The dataset we use is a 10-core subset, i.e. every user bought in total at least 10 items ($\sum_{B\in\mathscr{B}^u}|B|) > 10$ and vice versa each item was bought by at least 10 users. The statistics of the dataset can be found in table 8.1. We also created a dense subset of the 10-core dataset to study the effect of sparsity on the methods.

[1] `http://www.rossmannversand.de/`

Table 8.1 Characteristics of the datasets in our experiments in terms of number of users, items, baskets and triples (u,i,t) where t is the sequential time of the basket. The dense dataset is a subset of the sparse one containing the 10,000 users with most purchases and the 1000 items the most purchased.

Dataset	users	items	baskets	avg. basket size	avg. baskets / user	triples
Drug store (sparse)	71,602	7,180	233,476	11.3	3.2	2,635,125
Drug store (dense)	10,000	1,002	90,655	9.2	9.0	831,442

8.5.2 Evaluation Methodology

We evaluated by splitting the dataset S into two non overlapping sets: a training set S_{train} and a testing set S_{test}. This split is done by putting the last basket for each user into S_{test} and the remaining ones into S_{train}. The recommenders were trained on S_{train} and then the performance on S_{test} is measured. Hyperparameter search is done by removing for each user the last basket of S_{train} and using these baskets for the validation set.

Additionally, we removed those users from the evaluation that have bought less then 10 different items in the past (i.e. S_{train}). Secondly, for each user we removed all items from the test baskets (and the corresponding predictions) that this user has already bought in the past – this is because we want to recommend to the user items that are new/ unknown to him. Note that this makes the prediction task much harder and explains the low f-measure of all methods in figure 8.5. Otherwise just rerecommending already bought items would be a simple but very successful strategy for non-durable products in drug stores like toothbrushes or cleaner. However, this is not the task of recommender systems because they should help the user to discover new things.

The quality is measured for each user u on the basket B_u in the test dataset. We use the quality measures HLU, top-5 F-Measure and AUC to evaluate the estimated ranking against the actual bought items (see section 3.5). The runtime of learning the model parameters linearly depends on the number of factorization dimensions. With our implementation, training of the largest models ($k = 128$) took about 4 hours for MF, 31 hours for FMC and 34 hours for FPMC on the larger (sparse) dataset.

8.5.3 Results

In figure 8.5 you can see the quality on the sparse and dense online-shopping dataset. For the factorization methods we run each method with $k_{U,I} = k_{I,L} \in \{8, 16, 32, 64, 128\}$ factorization dimension. The x-axis of the diagrams reflects this increasing dimensionality. As expected all methods outperform the most-popular baseline clearly on both datasets and all quality measures. Secondly, with reasonable factorization dimensions (e.g. 32) all the factorization methods outperform the

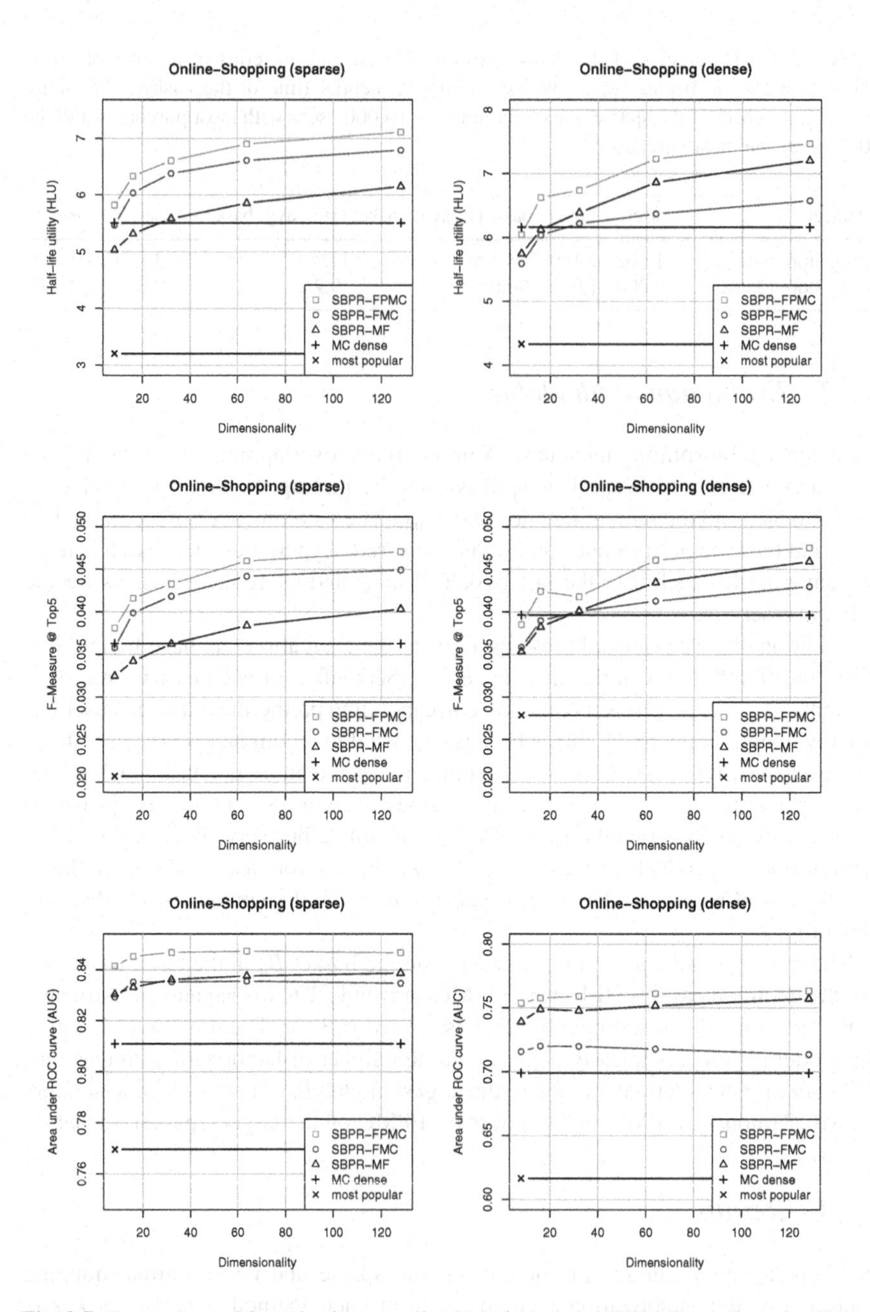

Fig. 8.5 Comparison of factorized personalized Markov chains (FPMC) to a factorized Markov chain (FMC), matrix factorization (MF) (Rendle et al, 2009), a standard dense Markov chain (MC dense) learned with Maximum Likelihood and the baseline 'most-popular'. The factorization dimensionality is increased from 8 to 128.

Table 8.2 Properties of the MC transition matrix estimated by the counting scheme. For the sparse dataset, only 12% of the entries of the transition matrix are non-zero and non-missing. For the dense subset, 88% are filled.

Dataset	total	missing values	non-zero	zero
Drug store (sparse)	51,552,400	1,041,100 (2.0%)	6,234,371 (12.1 %)	44,276,929 (85.9%)
Drug store (dense)	1,004.004	0 (0.0%)	889,419 (88.6 %)	114,585 (11.4%)

standard MC method. And in total, the factorized personalized MC (FPMC) outperforms all other methods.

8.5.3.1 MC vs. FMC

First, we want to discuss the advantage of factorization over a dense transition model by comparing MC with non-personalized FMC. The results indicate that learning a factorized transition matrix leads to better estimates than usual counting schemes. Factorization has two advantages (1) it can densify a sparse transition matrix and (2) it prevents overfitting of the estimates by using a low-rank approximation. The sparseness of the transition matrix estimated by counting schemes can be seen in table 8.2. In the dense setting also the transition matrix is filled in 88% whereas on the sparse dataset this drops to 12%. Comparing the quality on the sparse and dense setting in figure 8.5, one can see that the advantages of FMC over MC are much higher in the sparse setting than in the dense one. But even in the dense setting where also MC's transition matrix is almost completely filled, FMC outperforms MC because the factorization prevents overfitting by using less parameters.

8.5.3.2 MF vs. FMC vs. FPMC

Comparing the factorized Markov chain with the matrix factorization, one can see that in the dense setting MF seems to outperform MC whereas in the sparse one MC is superior. The reason could be that in the dense setting there is much more information per user, thus the MF method using all the users purchase information has advantages over the MC model that only relies on the last purchases. And the other way around, MC has advantages on the sparse dataset. FPMF that combines the advantages of both methods outperforms them on both datasets.

8.5.3.3 Dense vs. Sparse Data

On the dense data, all methods perform better with respect to HLU and top-5 F-Measure. That means that for dense data it is easier to rank a match high in the result list. On the other hand, for all methods the AUC is much higher on the sparse data.

The reason is that AUC evaluates all positions equally and is not restricted to the top positions. So, methods on the sparse dataset benefit from having many items that are very rarely selected and thus it is easy to rank them low and increase the AUC. In contrast to this the dense dataset contains only the 'hard' items (with more than 10 selections in train) and so the AUC is not increased by easy cases like in the sparse data.

8.6 Conclusion

We have introduced a recommender method based on personalized Markov chains over sequential set data. Instead of using the same transition matrix for all users, this method uses an individual transition matrix for each user which in total results in a transition cube. As direct estimation (e.g. by Maximum Likelihood) over a fully parametrized transition cube leads to very poor transitions, we introduce a factorization model that gives a low-rank approximation to the transition cube. The advantages of this approach is that each transition is influenced by transitions of similar users, similar items and similar transitions. Thus the quality of the final transition graph is much higher than that of a fully parametrized model. Secondly, we apply factorized personalized Markov chains (FPMC) to the task of item recommendation with sequential set data by adapting the BCR framework. Additionally, we show that FPMC subsumes the popular matrix factorization model and a non-personalized factorized Markov chain. Due to the expressiveness of FPMC it combines the advantages of both the state-of-the-art global personalized approach (MF) and the sequential MC method. Empirically, we show on real-world data that FPMC outperforms MF, FMC and normal MC both on sparse and dense data.

References

Hu, Y., Koren, Y., Volinsky, C.: Collaborative filtering for implicit feedback datasets. In: IEEE International Conference on Data Mining (ICDM 2008), pp. 263–272 (2008)

Koren, Y.: Factorization meets the neighborhood: a multifaceted collaborative filtering model. In: KDD 2008: Proceeding of the 14th ACM SIGKDD International Conference on Knowledge Discovery and Data Mining, pp. 426–434. ACM, New York (2008)

Koren, Y.: Collaborative filtering with temporal dynamics. In: KDD 2009: Proceedings of the 15th ACM SIGKDD International Conference on Knowledge Discovery and Data Mining, pp. 447–456. ACM, New York (2009)

Mobasher, B., Dai, H., Luo, T., Nakagawa, M.: Using sequential and non-sequential patterns in predictive web usage mining tasks. In: ICDM 2002: Proceedings of the 2002 IEEE International Conference on Data Mining, p. 669. IEEE Computer Society, Washington (2002)

Pan, R., Scholz, M.: Mind the gaps: weighting the unknown in large-scale one-class collaborative filtering. In: KDD 2009: Proceedings of the 15th ACM SIGKDD International Conference on Knowledge Discovery and Data Mining, pp. 667–676. ACM, New York (2009)

Rendle, S., Freudenthaler, C., Gantner, Z., Schmidt-Thieme, L.: BPR: Bayesian personalized ranking from implicit feedback. In: Proceedings of the 25th Conference on Uncertainty in Artificial Intelligence (UAI 2009) (2009)

Shani, G., Heckerman, D., Brafman, R.I.: An mdp-based recommender system. Journal of Machine Learning Research 6, 1265–1295 (2005)

Zimdars, A., Chickering, D.M., Meek, C.: Using temporal data for making recommendations. In: UAI 2001: Proceedings of the 17th Conference in Uncertainty in Artificial Intelligence, pp. 580–588. Morgan Kaufmann Publishers Inc., San Francisco (2001)

Part IV
Extensions

Finally, we investigate two extensions of our work. The first one is a general extension of factorization models where the factors are time-variant. The second one targets binary classification for two-mode settings where only one class is observed. Even though both work is related to context-aware ranking with factorization models, they are neither a direct extension nor an application. Instead the focus is on other aspects: (1) modelling time-aware factors in general and (2) solving balanced one-class problems with matrix factorization.

First, we deal with the problem of modelling time-variance within factorization models in general. Our approach is to make each factor (e.g. of the factorization matrix) time-dependent. The idea is to model the time-variant factor with a linear combination of a set of basis functions. This can be seen as factoring each time-variant factor itself into time-independent parameters and the basis functions. Instead of using a predefined set of basis functions, we will generate them from the data using a kernel approach. This way, most expressiveness in terms of time-variance is put into regions of time where most observations were made.

Whereas the first extension focuses on the model, the second extension deals with a special problem setting. The observed data in context-aware ranking can be seen as a one-class problem. For example in item recommendation, we only observe what a user has done – e.g. what products he buys. Then, the task is to predict what products he buys next. For all context-aware ranking problems, the class imbalance is huge – i.e. the products he will not buy largely dominate the products he will buy. Thus, the task is to rank/ recommend the products instead of classifying them. Here, we will deal with another problem setting, where the classes are more or less balanced. That means even though, only examples of one class are observed, on future test data the classes are balanced. To solve this issue, we will transfer ideas of one-class support-vector-machines to matrix factorization.

Chapter 9
Time-Variant Factorization Models

All of our proposed factorization models so far do not model any time-variance within factors. In this chapter, we develop a non-parametric approach that allows to model changes in time for each factor. We will focus on the model itself which is generic and not limited to any optimization task like ranking, classification or regression. Even though, factor models for context-aware ranking can benefit from these extensions, this work is not limited to the ranking task, but is more general. That is why we describe the time-aware factorization models for typical problems instead of limiting the discussion to context-aware ranking.

9.1 Introduction

Factorization models (FM) are the basis for many popular methods in machine learning, including maximum margin matrix factorization (Srebro et al, 2005), (higher-order) singular value decomposition (Lathauwer et al, 2000) or principal component analysis. They subsume many models like Tucker decomposition (Tucker, 1966), Parallel factor analysis/ PARAFAC (Harshman, 1970; Carroll and Chang, 1970) or matrix factorization. The usual assumption for factorization models is that the factors are independent of time. In this chapter, we extend general FM to the case where factors are time-variant.

Analogously to the general idea of factorization models, we decompose each time-variant factor into a time independent part and a time-variant part. The time-variant part is a set of basis functions that is modelled explicitly. Instead of using the same predefined set of time-variant functions for each factor, we generate these functions from the data using a non-parametric kernel approach. This is done per factor because in factorization problems among the variables the time points with (latent) observations differ and also the number of observations differ. Our generated functions make sure that the free model parameters that have to be estimated are placed corresponding to the time points with observations of this variable – e.g. regions with more observations get more parameters. This leads both to a closer fit and less overfitting because the number of basis functions can be reduced and related to the number of observations.

S. Rendle: Context-Aware Ranking with Factorization Models, SCI 330, pp. 137–153.
springerlink.com

First, we shortly introduce the general problem of modelling relations over high-dimensional categorical domains with factorization models. Then, we introduce our time-variant factorization model and show the decomposition of the time-variant factors into free parameters and a set of basis functions. Then we develop a general probabilistic model for this decomposition. This is used to derive simple basis functions like time-invariant biases and single point estimates. After that we introduce our generating approach that is based on a probabilistic kernel between two points of time. Using this kernel the time-variant basis functions can be generated by Gibbs sampling. Finally, we evaluate this approach on a real world dataset and four artificial datasets with varying characteristics.

Besides the rich literature on time-independent factorization models, there is only little research on time-variant settings. Recently, a time-variant matrix factorization model for collaborative filtering has been introduced (Koren, 2009). Koren investigates empirically several time models including bins, drift and a spline[1]. In contrast to this, we introduce a generic model for factorization models (including higher-order $m > 2$) and develop a method for generating the basis functions from the data.

9.2 Problem

Factorization models can model relations over categorical domains. For instance, the matrix factorization model in chapter 6 models the interactions between users and items – e.g. how much a user likes an item. It has been shown empirically (Koren, 2009) that this problem is time-variant and that e.g. a user's taste changes over time.

9.2.1 Time-Variant Relations

We use the notation of chapter 3: Let $X_1, \ldots, X_m$ be the domains of m categorical variables, i.e. $X_i = \{x^i_1, \ldots, x^i_{|X_i|}\}$. Depending on the problem, variables with the same domain are possible – e.g. for representing symmetric matrices like adjacency graphs.

A time-variant relation Y over the categorical domains $X_1, \ldots, X_m$ can be represented as a function y:

$$y : X_1 \times \ldots \times X_m \times \mathbb{R} \to T \tag{9.1}$$

where T is the target domain, e.g. $T \subseteq \mathbb{R}$ for regression problems or $T = \{-1, 1\}$ for classification. Or alternatively in tensor notation:

$$Y : \mathbb{R} \to T^{|X_1| \times \ldots \times |X_m|} \tag{9.2}$$

We will use both notations $y_{x_1,\ldots,x_m}(t)$ and $y(x_1, \ldots, x_m, t)$.

[1] This spline approach is related to our kernel method using a special kernel and sampling independently of the observations.

9.2.2 Sparseness

We assume that only a small part $S \subset (X_1 \times \ldots X_m \times \mathbb{R})$ of the relation Y is observed. Sparseness is assumed both in the dimensions of the variables and even more in time. That means for each variable instance $x \in X$ there are only limited points in time where the relation Y is observed. These are the points $T_x := \{t | \hat{f}_T^x(t) > 0\}$ where the empirical distribution $\hat{f}_T^x$ is non-zero:

$$\hat{f}_T^x(t) := \frac{|(X_1 \times \ldots \{x\} \times \ldots X_m \times \{t\}) \cap S|}{|(X_1 \times \ldots \{x\} \times \ldots X_m \times \mathbb{R}) \cap S|} \tag{9.3}$$

We assume that each variable instance x is observed at least once and thus the denominator does not vanish. Furthermore, the empirical distribution over a relation instance $\mathbf{x} = (x_1, \ldots, x_m)$ has only very small or no support as most relation instances are never observed. Note that this makes the prediction task very hard, because it means that time-variance is not observed directly on the relation instances but is hidden within variable interactions. Thus the problem is more difficult than typical time series problems.

9.2.3 Context-Aware Ranking

Now, we will briefly show how context-aware ranking fits into this setting.

Problem Setting

For ranking within a time-aware problem (eq. (9.1)), the context would be:

$$\mathscr{C} = X_1 \times \ldots \times X_{m-1} \times \mathbb{R} \tag{9.4}$$

And the target is to find a ranking on X_m given a context $\mathbf{c} \in \mathscr{C}$:

$$\succ \subset \mathscr{C} \times X_m^2 \tag{9.5}$$

Modelling

Like discussed in section 3.4, the ranking can be expressed by the function y (eq. (9.1)). But now, y contains a variable (the time) that is not finite and thus the factorization models of chapter 5 cannot be applied directly. Thus, in this chapter we derive time-aware factorization models that can handle time. The rest of this chapter focuses on this task.

Learning

The learning task is then to find a function y that generates a ranking $\succ^y$ which satisfies the infered training data D_S. E.g. by using BCR-OPT and BCR-LEARN.

Sparseness

The semantics of S for the general case (see section 9.2.2) is different from the training data in context-aware ranking. The reason is, that for context-aware ranking there are no direct observations of y itself. Instead, training data is given on pairs within D_S which are direct training data for $\succ$ and thus D_S is indirect training data for y. Nevertheless, we will never need S from now on, but only $\hat{f}_T^x$. This can be defined for context-aware ranking as:

$$\hat{f}_T^x(t) := \begin{cases} \frac{|(\mathscr{C} \times \{x\} \times X_m \times \{t\}) \cap D_S| + |(\mathscr{C} \times X_m \times \{x\} \times \{t\}) \cap D_S|}{|(\mathscr{C} \times \{x\} \times X_m \times \mathbb{R}) \cap D_S| + |(\mathscr{C} \times X_m \times \{x\} \times \mathbb{R}) \cap D_S|} & \text{if } x \in X_m \\ \frac{|(X_1 \times \ldots \{x\} \times \ldots X_m^2 \times \{t\}) \cap D_S|}{|(X_1 \times \ldots \{x\} \times \ldots X_m^2 \times \mathbb{R}) \cap D_S|} & \text{else} \end{cases} \tag{9.6}$$

With these definitions, also context-aware ranking can be applied to the general model on which we will focus from now on.

9.3 Time-Variant Factorization Models

In the following, we discuss approaches to model $Y(t)$ by a factorization model $\hat{Y}(t)$. First, we introduce the time-variant factorization models and derive a general way of modelling time-variant factors by decomposing them into latent parameters Θ and a predefined time-structure $\mathscr{H}(t)$.

9.3.1 Time-Variant Tucker Decomposition

Based on the example of Tucker decomposition (TD), we introduce time-variant factor models. The TD model of a tensor of mode m is defined as:

$$\hat{Y} := \mathscr{B} \times_1 V^1 \times_2 \ldots \times_m V^m \tag{9.7}$$

$$\mathscr{B} \in \mathbb{R}^{k_1,\ldots,k_m}, \quad V^i \in \mathbb{R}^{|X_i| \times k_i} \tag{9.8}$$

Where V^i is the factorization matrix for a variable X_i. For each variable instance $x \in X_i$, V^i contains one row with k_i values that describe the factors of x. It is very important to note that these factors are never observed, but have to be estimated during the learning phase of the model. TD subsumes a variety of factorization models including PARAFAC or matrix factorization. See chapter 5 for more details about TD and PARAFAC.

A time-variant TD is modelled by making the factors depending on time:

$$\hat{Y}(t) := \mathscr{B}(t) \times_1 V^1(t) \times_2 \ldots \times_m V^m(t) \tag{9.9}$$

$$\mathscr{B} : \mathbb{R} \to \mathbb{R}^{k_1,\ldots,k_m}, \quad V^i : \mathbb{R} \to \mathbb{R}^{|X_i| \times k_i} \tag{9.10}$$

The core tensor $\mathscr{B}$ is usually chosen fixed (e.g. diagonal for PARAFAC/ MF) or for TD it can be computed directly from orthonormal factorization matrices. Thus, we will concentrate on modelling time variant factorization matrices $V^i(t)$. As the modelling is identical for all $i \in \{1,\ldots,m\}$ and to shorten notation, we will drop the index and write $V(t)$ and X respectively. Recall that neither the factors nor the time-variance on variable instances is ever directly observed.

9.3.2 Time-Variant Factor Matrices

Let V be a factorization matrix for a variable X. The task is to model changes in time of $V(t)$. Our idea is to decompose $V(t)$ into value estimates Θ and general time-variant functions $\mathscr{H}(t)$.

$$\hat{V}(t) := \Theta \otimes^* \mathscr{H}(t) \tag{9.11}$$

where Θ is a parameter tensor:

$$\Theta \in \mathbb{R}^{|X| \times k \times l} \tag{9.12}$$

and $\mathscr{H}$ contains l basis functions that are assumed to describe the general-time dependencies of the problem. In general, these time dependencies can be individual per variable instance and factorization feature:

$$\mathscr{H}: \quad \mathbb{R} \to \mathbb{R}^{|X| \times k \times l} \tag{9.13}$$

And $\otimes^*$ is defined as the multiplication and contraction operation that multiplies entries with identical index of X and k and contracts l:

$$\hat{v}_{x,f}(t) := \sum_{i=1}^{l} h_{x,f,i}(t)\,\theta_{x,f,i} \tag{9.14}$$

In this model, Θ are the latent parameters to be estimated and $\mathscr{H}$ is a predefined tensor of functions that is modelled explicitly. Thus with this approach, the task of modelling $V(t)$ is reduced to model $\mathscr{H}(t)$. In the following, we will describe how to obtain models for each factor $v_{x,f}(t) \in V(t)$ – i.e. by defining the basis functions $h_{x,f,i}$ in the set $H_{x,f} := \{h_{x,f,1},\ldots,h_{x,f,l}\}$. To shorten notation, we will skip all indices x,f whenever possible and write $v(t)$.

9.4 Models for Time-Variant Factors

First, we investigate the general model for $v(t)$ that is described by a set H of time-variant basis functions. Next, we derive a probabilistic model that can express any bounded factor. Based on the assumption that the values of a factor at two similar points should be similar, we develop a non-parametric kernel based model that generates the functions in H based on the observed data in S.

9.4.1 General Decomposition Model

The general model for a time-variant factor is:

$$\hat{v}(t) := \sum_{h \in H} h(t)\,\theta_h \tag{9.15}$$

Where H is the set of basis functions for modelling the factor v.

Expressiveness

In the general case, H is not restricted:

$$H^{\text{general}} \subseteq \{h : \mathbb{R} \rightarrow \mathbb{R}\} \tag{9.16}$$

Obviously, without a restriction on H, the model in eq. (9.15) can express any function $v(t)$. This can easily be seen by:

$$H := \{v(t)\}, \quad \theta_1 := 1 \quad \Rightarrow \hat{v}(t) = v(t) \tag{9.17}$$

For sure this shows only the theoretical expressiveness as v is not known in practice.

Modelling Approaches

An example for H is the set of trigonometric functions:

$$H^{\text{trig}} := \{\sin(nt), \cos(nt) \,|\, n \in \mathbb{N}\} \tag{9.18}$$

with this eq. (9.15) becomes a fourier series which is known to approximate any function well within a fixed interval. Our approach is different from that. Rather than taking the same fixed set of functions for all factors, we generate the functions from the data. To derive these functions, we first formulate a probabilistic model.

9.4.2 Probabilistic Model

For the probabilistic analysis we use the following random variables:

- $T \in \mathbb{R}$ is the random variable for time – t is the realization.
- $V \in \mathbb{R}$ is the random variable of the factor – v is the realization.
- $C \in \{c_1, \ldots, c_l\}$ is the random variable over a finite set of features for explaining the time behaviour of V.

Keep in mind, that v is a shortcut for $v_{x,f}$ and also C is depending on (x, f) – i.e. $C_{x,f}$.

9.4.2.1 General Model for Bounded Factors

Similar to the algebraic decomposition of value estimation and time, we assume conditional independence between factor V and time T given C that explains the time variant behaviour:

$$p(v,t|c) \stackrel{!}{=} p(v|c)\,p(t|c) \tag{9.19}$$

We will show in the following how to relate this model to the general model in eq. (9.15). Afterwards, our kernel model will be a specialization of this general probabilistic model.

Expectation Value $E(v|t)$

It is well known that the (conditional) expectation $E(v|t)$ of a random variate is minimizing the squared prediction error for any variable v at a predetermined point of time t. Hence, a straightforward way of modelling the time-variant factor is:

$$\hat{v}(t) := E(v|t) \tag{9.20}$$

This expectation value can be expressed as a function of C.

Lemma 9.1. *With eq. (9.19), the expected value for v at time point t is:*

$$E(v|t) = \sum_{c\in C} E(v|c)\,p(c|t) \tag{9.21}$$

Proof. With the definition of expectation we have:

$$E(v|t) := \int_{\mathbb{R}} v\,p(v|t)\,dv \tag{9.22}$$

Next, $p(v|t)$ is transformed by marginalization with C:

$$\begin{aligned} p(v|t) &= \sum_{c\in C} p(v,c|t) = \sum_{c\in C} \frac{p(v,t|c)\,p(c)}{p(t)} \\ &\stackrel{(*)}{=} \sum_{c\in C} \frac{p(v|c)\,p(t|c)\,p(c)}{p(t)} = \sum_{c\in C} p(v|c)\,p(c|t) \end{aligned} \tag{9.23}$$

At $(*)$ the conditional independence of v and t given the time-effect variable c is used. Substituting $p(v|t)$ in the expectation leads to:

$$\begin{aligned} E(v|t) &= \int_{\mathbb{R}} v \sum_{c\in C} p(v|c)\,p(c|t)\,dv \\ &= \sum_{c\in C} p(c|t) \underbrace{\int_{\mathbb{R}} v\,p(v|c)\,dv}_{E(v|c)} = \sum_{c\in C} E(v|c)\,p(c|t) \end{aligned} \tag{9.24}$$

Probabilistic Model

In total, the probabilistic model for the factor v is:

$$\hat{v}(t) = \sum_{c \in C} E(v|c)\, p(c|t) \tag{9.25}$$

As discussed before, no parameter in our latent model is directly observed, so $\theta_c := E(v|c)$ has to be estimated in the model's learning phase:

$$\hat{v}(t) = \sum_{c \in C} \theta_c\, p(c|t) \tag{9.26}$$

Comparing this to eq. (9.15) makes the semantics clear: the variables C with the distribution $p_{C|T}$ correspond to H and θ_h is the expectation value $E(v|c)$. Thus, given $p_{C|T}$, the function set H is defined as:

$$H^{\text{prob}}(p) := \{p(c_1|t), \ldots, p(c_l|t)\} \tag{9.27}$$

Expressiveness

With the probabilistic model, H is restricted to all conditional probability distributions over C:

$$H^{\text{prob}} \subseteq \left\{ p \,\middle|\, \forall t : \sum_{c \in C} p(c|t) = 1, \forall c : p(c|t) \geq 0 \right\}$$

This model can express any time variant factor $v(t)$ with a finite lower ($\eta_l := \min_t v(t)$) and upper bound ($\eta_u := \max_t v(t)$). This can be seen by using the following example.

$$\begin{aligned}
C &:= \{c_1, c_2, c_3\} \\
p(c_1|t) &:= \frac{0.5}{\eta_u - \eta_l}(v(t) - \eta_l), & \theta_1 &:= 2(\eta_u - \eta_l) \\
p(c_2|t) &:= 0.5 - p(c_1|t), & \theta_2 &:= 0 \\
p(c_3|t) &:= 0.5, & \theta_3 &:= 2\eta_l
\end{aligned}$$

Now p is a probability distribution over C for each t and $\hat{v}(t) = v(t)$. Again, this only shows the theoretical expressiveness of this model.

9.4.2.2 Basic Probabilistic Models

In the following, we derive concrete models for $p_{C|T}$. To distinguish the distributions, we will use their random variables as index (e.g. $p_{C|T}$) and use f for densities and p distributions over discrete variables.

Single Point Estimates

Recall, that for each variable instance $x \in X$ there are only limited points in time where the relation Y is observed (see eq. (9.3)). At these time points $t \in T_x$ the latent variable $v_{x,f}(t)$ can be estimated. Using one variable c for each time point $t \in T_x$ a point estimate for $v(t)$ can be made:

$$|C| := |T_x|, \quad \phi : C \to T_x, \quad \phi \text{ bijective} \tag{9.28}$$

$$\forall c \in C : \; p(c|t) := \delta(\phi(c) = t) \tag{9.29}$$

where ϕ is the placement of the basis function δ via C on the time domain. Applying this to equation (9.26) leads to:

$$v(t) = \delta(t \in T_x)\, \theta_{\phi^{-1}(t)} \tag{9.30}$$

That means for each time with an observation of x, a time depending latent feature is learned such that $\hat{y}$ (see eq. (9.1)) can be reconstructed at that point of time. But (i) no estimations for non observed points of time for this variable can be made (besides the trivial estimate 0). Especially forecasting is not possible. And (ii) no relationships between the values of v at two close points of time are made. Thus, single point estimates might suffer from overfitting. In general this model is useful to model noise on the observations.

Constant effects/ Bias

Often, the largest part of a latent feature is time-independent – i.e. the time-variance is centered around a fixed bias. Modelling time-independence is done by:

$$C := \{c\}, \quad p(c|t) := 1 \tag{9.31}$$

Mixture effects

Mostly, an effect does not appear isolated but mixed with other effects. Two effects $(C_1, p(C_1|t))$ and $(C_2, p(C_2|t))$ can be mixed by linearly combining both into a new effect $(C, p(C|t))$:

$$C := C_1 \cup C_2 \tag{9.32}$$

$$\forall c_i \in C : \; p(c_i|t) := \begin{cases} \alpha\, p(c_i|t), & \text{if } c_i \in C_1 \\ (1-\alpha)\, p(c_i|t), & \text{if } c_i \in C_2 \end{cases}$$

where $\alpha \in [0,1]$ defines the weight for the combination. For sure, also more than two effects can be mixed – e.g. by recursively applying eq. (9.32). For example a reasonable mixture is a combination of a bias, noise (point estimates) and our non-parametric model that we describe next.

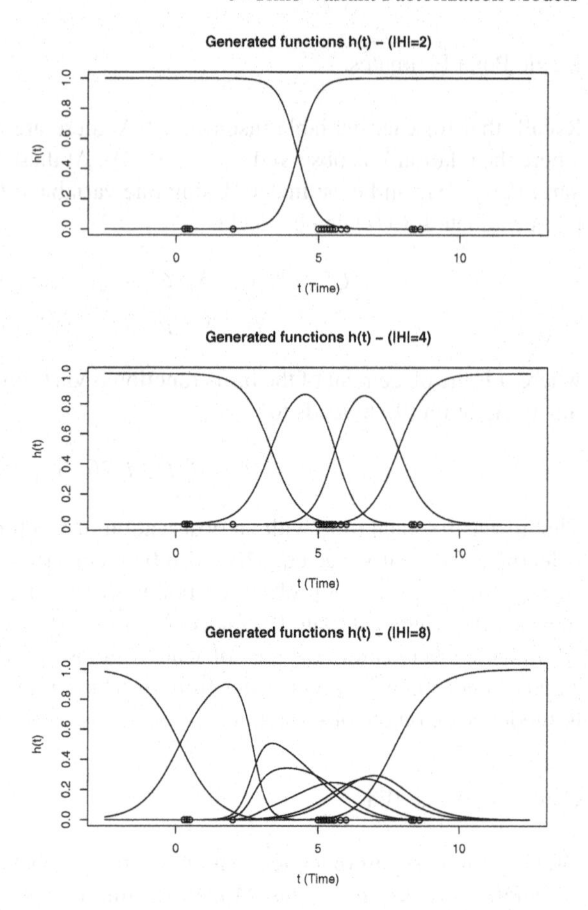

Fig. 9.1 Basis functions H_v for a variable instance x generated by our non-parametric method. This method is based on a kernel function (here Gaussian kernel) and the observed time points of this variable instance. Here, these time points are plotted as dots on the time-axis. As the factor v should be approximated by $|H|$ basis functions, more complexity is placed to regions of time with more (indirect) observations.

9.4.3 Non-parametric Method for Generating Time-Variant Basis Functions

In this section, we derive a method to generate for each variable instance x continuous functions H_x from the observed data S. The model is based on the idea of a continuous dependency K of two points of time – we call this dependency the 'kernel'. We will present how to model $p_{C|T} \cong H$ only with this kernel assumption using the data to 'place' the variables C in regions with many observations (see figure 9.1).

Continuous Time Dependency

The time-dependency of two points of time t_1 and t_2 is given as the density $K(\alpha)$ over their difference $\alpha = t_1 - t_2$. We denote this time dependency as K because we will refer to it as a (time) *kernel*.

Examples for time kernels are:

$$K^{\text{Gauss}}(\alpha) := \frac{1}{\sqrt{2\pi\sigma^2}} \exp\left(-\frac{\alpha^2}{2\sigma^2}\right) \tag{9.33}$$

$$K^{\text{Exp}}(\alpha) := \delta(\alpha > 0)\,\lambda e^{-\lambda\,\alpha} \tag{9.34}$$

$$K^{\text{Identity}}(\alpha) := \delta(\alpha = 0) \tag{9.35}$$

After a kernel has been chosen, it can be used to define the conditionals between time t and the latent time variables C[2]:

$$f_{T|C}(t|c) := K(t - \phi(c)) \tag{9.36}$$

Again, a mapping $\phi : C \to \mathbb{R}$ from time effect variables to the time domain is necessary. We will later show how to obtain this mapping by Gibbs sampling. With this definition, the exponential kernel can be seen as a causal kernel, because the support of K is only in the future of $\phi(c)$ – that means c influences only the future. E.g. Poisson processes have the exponential kernel as density. The identity kernel leads to single point estimates as described in eq. (9.30).

Kernel-based Model for $p_{C|T}$

We can express $p_{C|T}$ by $f_{T|C}$ using the Bayes theorem:

$$p_{C|T}(c|t) = \frac{f_{T|C}(t|c)\,p_C(c)}{\sum_{c' \in C} f_{T|C}(t|c')\,p_C(c')} \tag{9.37}$$

Using the kernel definition and assuming p_C to be uniformly distributed results in:

$$p_{C|T}(c|t) = \frac{1}{Z(t)} K(t - \phi(c)) \tag{9.38}$$

$$\text{with} \quad Z(t) := \sum_{c \in C} K(t - \phi(c))$$

which defines the basic functions H (eq. (9.27)). An example can be found in figure 9.2. Here we have three variables that have been placed on time points $2, 4, 7$. For each time t on the t-axis, the y-axis shows the probabilistic influence $p_{C|T}(c|t)$ (i.e. the weight) of each variable c on $v(t)$. In this figure, the basis functions $h_1(t) := p(c_1|t)$, $h_2(t) := p(c_2|t)$ and $h_3(t) := p(c_3|t)$ are stacked.

Placement ϕ of Latent Variables

Next, we show how to derive the placement of C on the time domain. It is based on sampling $|C|$ points of time using the empirical distribution over time $\hat{f}_T$ and

---

[2] As C is a finite set and K is usually a density, one cannot use K directly to express $p_{C|T}$

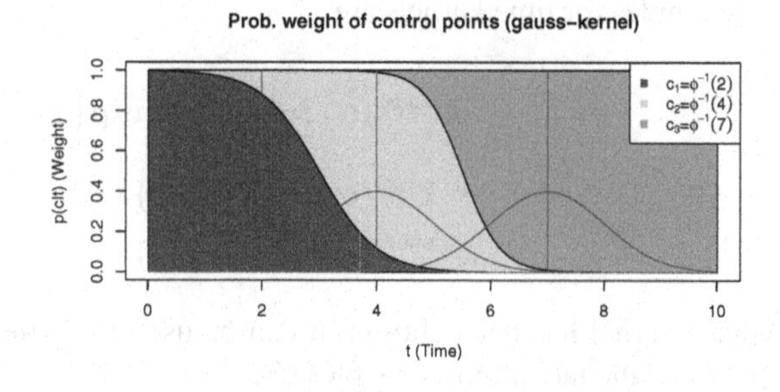

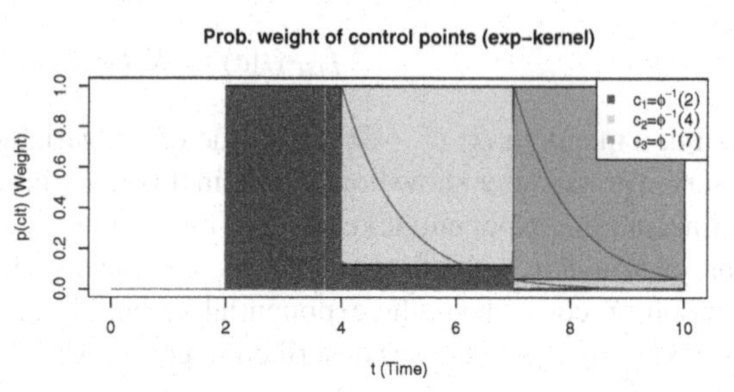

Fig. 9.2 Relationship between the kernel K and $p_{C|T}$ (which defines the basis functions H). There are three time variables c_1, c_2, c_3 placed at time $2, 4, 7$ with their corresponding kernel. The two figures show $p_{C|T}$ for the Gaussian kernel (*top*) and for an exponential kernel (*bottom*). The shaded areas show the weight of the time variables c_1, c_2, c_3. At all points of time $p_{C|T}$ is a probability distribution over C. An example for generated functions H is shown in figure 9.1.

the kernel. For each variable c, one sample t_c is drawn that defines the target of the mapping $\phi(c) := t_c$.

Let T_C be the continuous random variable to describe the placement of C on the time domain. By marginalization, one can express T_C by:

$$\begin{aligned} f_{T_C}(t_c) &= \int_{\mathbb{R}} f_{T_C,T}(t_c, t)\, dt \\ &= \int_{\mathbb{R}} f_{T_C|T}(t_c|t)\, f_T(t)\, dt \end{aligned} \tag{9.39}$$

Sampling directly from the joint probability distribution $f_{T_C,T}$ is not possible as it is unknown. Instead sampling from eq. (9.39) is possible, because the probability of f_T is empirically observed by $\hat{f}_T$ (eq. (9.3)) and the conditional $f_{T_C|T}$ is defined by the kernel K according to eq. (9.36). Thus:

$$f_{T_C}(t_c) \approx \int_{\mathbb{R}} K(t - t_c)\, \hat{f}_T(t)\, dt \tag{9.40}$$

Sampling from f_{T_C} can be performed by Markov Chain Monte Carlo methods (MCMC), i.e. a Gibbs sampler: (1) Sample t^* from $\hat{f}_T$; (2) Sample t_c from $K(t^* - t_c)$. Note that with eq. (9.38) and eq. (9.27), this corresponds to sampling functions h.

Generation Method

The complete procedure for generating for each variable instance x individual basis functions H_x is:

1. Choose a kernel function K.
2. Sample the mapping $\phi : C \rightarrow T$ with eq. (9.40)
3. The basis functions H_x are defined by eq. (9.38) and eq. (9.27).

This approach automatically places most parameters Θ to the regions with most observations $\hat{f}_T^x$ (see figure 9.1). The sampling process adapts automatically to the kernel – e.g. with an identity kernel the Gibbs sampler corresponds to bootstrapping. Furthermore, it is possible to use less parameters Θ for variables with only little observations by sampling less functions[3]. This leads to a better generalization capability and prevents overfitting.

The limitation of the kernel approach is that it does not infer from patterns in the past to patterns in the future. E.g. if a sine-like curve has been learned within the region of observations, this shape is not applied to the future. Instead, because of the model assumption that values are similar to close points, the curve would converge to the estimation θ of the last parameter – under the assumption of a decreasing kernel like exponential or Gaussian; another kernel e.g. a damped sine-kernel would behave differently and might correctly repeat the sine shape.

Expressiveness and Relation to ARIMA

If the kernel function can be chosen for each factor freely, because of eq. (9.37) the expressiveness is the same as the general probabilistic model in eq. (9.26).

If we limit the kernel to a fixed one, like Gaussian or exponential, it is still possible to keep the expressiveness of an arbitrary $p_{C|T}$ as long as we use an infinite number of parameters C and allow a free placement ϕ. For this proof, we quantise time $t = i\Delta t$ ($i \in \mathbb{Z}$, $\phi : C \rightarrow \mathbb{Z}$) and get from eq. (9.26) and eq. (9.38):

$$v(i\Delta t) = \sum_{c \in C} \theta_c \, g(i\Delta t - \phi(c)\Delta t) \tag{9.41}$$

with:

$$g(i\Delta t - \phi(c)\Delta t) := \frac{1}{Z(i\Delta t)} K(i\Delta t - \phi(c)\Delta t) \tag{9.42}$$

Applying the Z-transform and using its property of linearity and time shifting, this equation becomes:

$$v(z) = g(z)\,q(z) \tag{9.43}$$

$$q(z) = \sum_{c \in C} \theta_c \, z^{-\phi(c)\Delta t} \tag{9.44}$$

Eq. (9.44) is a Laurent-series which converges to a holomorphic function on (a possibly empty) annulus around 0. Any function holomorphic on an annulus has a

[3] In our evaluation, we sample $\#_x^\gamma$ times, where $\gamma \in [0,1]$ is a hyperparameter that controls the number of draws and $\#_x$ is the number of observations for the variable instance x.

convergent Laurent-series with respect to this annulus. Thus any (discrete) time dependency such that the Z-transform is holomorphic on an annulus can be represented by a sum (9.41) with infinitely countable C. In this sense, a model of this type is general enough.

A popular and generic time dependency of an output $v(t)$ on its input is the Box-Jenkins model (Ljung, 1999).

$$v(z) = \frac{c(z)/d(z)}{1 - a(z)/b(z)} g(z) \tag{9.45}$$

where $a(z), b(z), c(z), b(z)$ are given by Laurent series, and thus $\frac{c(z)/d(z)}{1-a(z)/b(z)}$ is rational. This model widely subsumes popular time series models including AR, ARMA and ARIMA. Because any rational function is holomorphic, $h(z) = \frac{c(z)/d(z)}{1-a(z)/b(z)}$ holds, and we conclude that our model covers all major time dependency of $v(t)$ as far as we allow C infinite.

9.5 Evaluation

We study the effectiveness of our method on regression problems – i.e. the square loss with L2-regularization is minimized:

$$\underset{\Theta_1,\ldots,\Theta_m}{\operatorname{argmin}} \sum_{(x_1,\ldots,x_m,t)\in S} (y_{x_1,\ldots,x_m}(t) - \hat{y}_{x_1,\ldots,x_m}(t))^2 + \Lambda \left(\sum_{i=1}^{m} ||\Theta_i||_F^2 \right) \tag{9.46}$$

We optimize this function by stochastic gradient descent which is known to work well on typical factorization problems (Koren, 2009). As factorization model we use PARAFAC which corresponds to the popular MF for cases with more than two modes.

9.5.1 Experimental Setup

9.5.1.1 Datasets

To get a deeper understanding of the effectiveness of our model, we use both synthetic and real-world data. The synthetic data is generated by the following PARAFAC-model.

$$y_{x_1,\ldots,x_m}(t) := \sum_{f=1}^{k} \prod_{i=1}^{m} v^i_{x_i,f}(t) + \varepsilon_{x_1,\ldots,x_m}(t) \tag{9.47}$$

Where $\varepsilon_{x_1,\ldots,x_m}(t) \sim N(0, 0.1)$ is Gaussian noise. To prevent outliers from dominating the evaluation, all values of y are truncated to $[-5, 5]$ – i.e. we cut the tails of the output distribution. Each of the factors $v^i_{x_i,f}$ of the generating model has its own time variance. We consider two types of time variances:

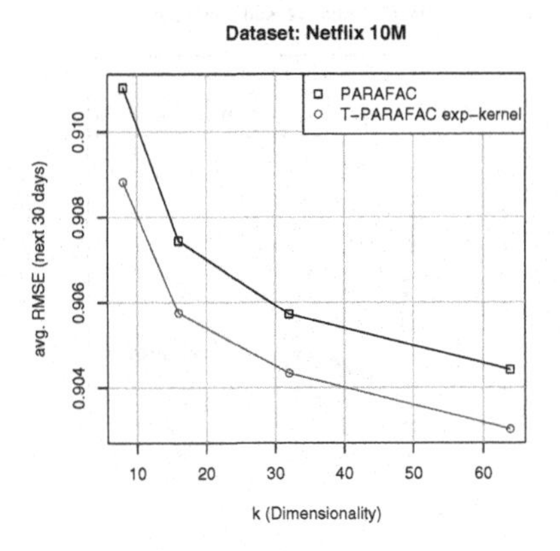

Fig. 9.3 Netflix 10M dataset: PARAFAC vs. T-PARAFAC (exp-kernel) with an increasing number of factorization dimensions (k). The evaluation was made for the last year in the dataset: for each month one model is learned on all past data and the next month is evaluated. The figure shows the average of these 12 RMSE scores per model type and dimensionality.

- **Stationary:** The generating functions are sines.

$$v^i_{x,f} := a^i_{x,f} + b^i_{x,f} \sin\left(\frac{2\pi c^i_{x,f} t}{t_{\max}} + d^i_{x,f}\right)$$

  Where $a^i_{x,f} \sim N(0,1)$, $b^i_{x,f} \sim N(0,0.5)$, $c^i_{x,f} \sim \text{Exp}(1)$, $d^i_{x,f} \sim N(0,\pi)$. Note that this corresponds to an ARMA model without damping, i.e. an ARMA model with strong memory and many parameters.
- **Trend:** The generating functions are linear trends:

$$v^i_{x,f} := a^i_{x,f} + b^i_{x,f} t$$

  Where $a^i_{x,f} \sim N(0,0.75)$ and $b^i_{x,f} \sim N(0,0.5/t_{\max})$.

We have generated 4 datasets: two 2-mode datasets with $|X_1| = |X_2| = 1000$ and two 3-mode datasets with $|X_1| = |X_2| = |X_3| = 100$. For each dataset $t_{\max} = 1000$ and we sampled $|S| = 100,000$ observations. The stationary models have $k = 4$ and the trend models $k = 8$ factors.

As real-world dataset we use a subset of the Netflix dataset with $|S| \approx 10,000,000$, $|X_{\text{user}}| = 40,000$ and $|X_{\text{item}}| = 5000$.

9.5.1.2 Evaluation Protocol

A forecasting problem is set up by splitting the datasets by time t_s. All data before t_s (i.e. $[-\infty, t_s)$) is put into the training set S. And the data S_{test} of the time-span $[t_s, t_s + 30)$ is used for evaluation. The model is trained on S and the RMSE on S_{test} is measured. By moving t_s with a step length of 30, we have independent test sets and overlapping training sets of increasing size. For the artificial datasets we increased t_s from 30 to 960 with the step length of 30. For the Netflix dataset we evaluated the last 12 months starting from day 1842.

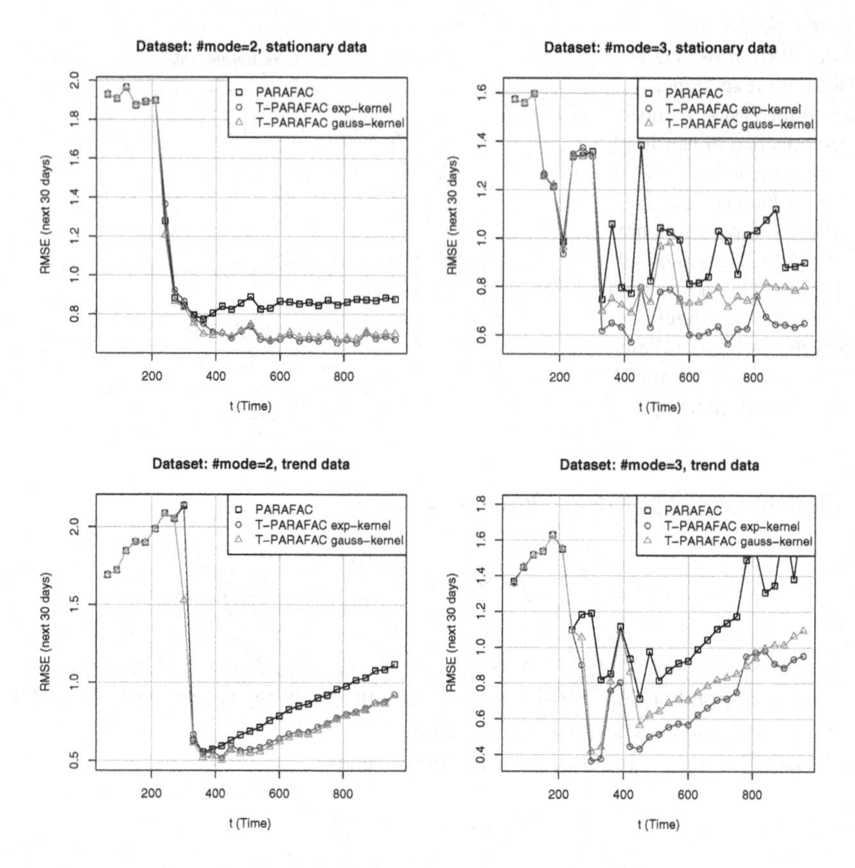

Fig. 9.4 Artificial data: Models are trained on a history from time 0 to time t. With increasing training data, T-PARAFAC (gaussian and exponential kernel) outperforms the standard PARAFAC model.

For the artificial datasets we run the models with several combinations of hyperparameters (k was fixed to the dimensions of the generating process) and the model selection at time t was to choose the hyperparameter combination that was best in the previous one[4]. For Netflix, the hyperparameters were chosen just once per holdout at the first time-split.

9.5.2 Results

Figure 9.3 compares on the Netflix dataset the average forecasting RMSE for the PARAFAC and time-variant PARAFAC (T-PARAFAC) model by varying the factorization dimensionality. As you can see, T-PARAFAC outperforms PARAFAC on

[4] Note that this corresponds to parameter selection using a holdout split of the previous 30 days.

all dimensionalities. This observation matches with previous results on this dataset (Koren, 2009).

Figure 9.4 shows how the next-month forecasting RMSE develops over time under heavily time-variant data. On all datasets, the models need about 250 time steps before enough data is accumulated to identify the factors. After that the T-PARAFAC outperforms PARAFAC. On the stationary data, increasing data helps the T-PARAFAC models to get more stable features whereas normal PARAFAC alternates. On trend data, after identifying the factors, all models suffer from more data. Still, T-PARAFAC has a better performance than PARAFAC as it seems to capture some of the time dependencies. We assume, that the reason why T-PARAFAC cannot fully capture the trend is the limited amount of data and that the trend is not explicitly modelled because the assumption of our model is that the future factors are similar to the last ones.

9.6 Conclusion

In this chapter, we have introduced a general model for time-variant factorization. The general model relies on decomposing a time-variant factor into a set of basis functions and time-invariant parameters. Based on the assumption that at two close points of time, the values of the factor are close, we have developed a method for generating the basis function for each variable instance individually. The generating process is based on a kernel that defines closeness and the time points of observations of the variable instance. We have discussed the expressiveness of this approach and the relations to ARMA models. In future work, we want to investigate how to apply time-patterns learned within a factor to enhance extrapolation to regions without observations. Furthermore, we want to apply the time-aware factor model to context-aware ranking problems.

References

Carroll, J., Chang, J.: Analysis of individual differences in multidimensional scaling via an n-way generalization of eckart-young decomposition. Psychometrika 35, 283–319 (1970)

Harshman, R.A.: Foundations of the parafac procedure: models and conditions for an 'exploratory' multimodal factor analysis. UCLA Working Papers in Phonetics, 1–84 (1970)

Koren, Y.: Collaborative filtering with temporal dynamics. In: KDD 2009: Proceedings of the 15th ACM SIGKDD International Conference on Knowledge Discovery and Data Mining, pp. 447–456. ACM, New York (2009)

Lathauwer, L.D., Moor, B.D., Vandewalle, J.: A multilinear singular value decomposition. SIAM J. Matrix Anal. Appl. 21(4), 1253–1278 (2000)

Ljung, L. (ed.): System identification (2nd ed.): theory for the user. Prentice Hall PTR, Englewood Cliffs (1999)

Srebro, N., Rennie, J.D.M., Jaakola, T.S.: Maximum-margin matrix factorization. In: Advances in Neural Information Processing Systems, vol. 17, pp. 1329–1336. MIT Press, Cambridge (2005)

Tucker, L.: Some mathematical notes on three-mode factor analysis. Psychometrika 31, 279–311 (1966)

Chapter 10
One-Class Matrix Factorization

The second topic, we are investigating is binary classification with matrix factorization models where only observations of one class are available. But in contrast to the problem settings we have discussed so far (see chapter 3), we now deal with a binary classification problem where the classes are more or less balanced. Context-aware ranking like item recommendation (see chapter 6) differs from this because it is a ranking task. Even when seeing it as a binary classification task, the problem differs substantially because the classes are typically hugely imbalanced: e.g. a customer buys much less books than he does not buy, a user listens to much less songs than he never listens to, etc.

Matrix factorization (MF) methods – e.g. based on maximum-margin learning (MMMF (Srebro et al, 2005)) – are known to be one of the best approaches for classification over two categorical variables. But sometimes only observations of one class are monitored and available to the learning algorithm, e.g. a website monitors what a user likes but not what he dislikes. In this chapter, we develop a MF method for solving such tasks where two-class methods are not applicable any more. We transfer the ideas of 1C-SVM to one-class maximum margin MF (1C-MMMF) which means biasing the classification to the unobserved class. We proof that 1C-MMMF is invariant to the particular size of this bias and that 1C-MMMF and 2C-MMMF have identical objective criteria. To solve this, we introduce a second regularizer based on the class prior of the unlabeled data, e.g. the test set. Our experiments indicate that our method outperforms the other approaches in quality and scales to problem sizes that are infeasible for 1C-SVM.

10.1 Introduction

Matrix factorization methods (MF) are well-studied for the task of predicting values of a binary relation over two categorical variables. They can solve both classification and regression problems of large scale with a high dimensionality of the two variable domains (Rennie and Srebro, 2005). In this chapter, we study the classification setting where only one class is observed and propose a one-class approach for matrix factorization models.

S. Rendle: Context-Aware Ranking with Factorization Models, SCI 330, pp. 155–170.
springerlink.com

In general, 2C-MMMF can be seen as the counterpart of linear 2C-SVM for the task of classification over two categorical domains because both are maximum-margin classifiers with soft margins as can be seen from their objective criteria. As SVMs have already been adapted to one-class problems (Schölkopf et al, 2001), we first transfer 1C-SVM to 1C-MMMF. The idea of 1C-SVM is to bias the prediction to the non-observed class and to optimize simultaneously this bias and the classification loss under maximum margin. We show that 1C-MMMF is invariant to the size/length of this bias and the optimization (but not the prediction) is equivalent to 2C-MMMF. Hence, 1C-MMMF is unable to identify the factorization dimensions because only positive examples are used during learning. To solve this, we introduce an additional regularizer on the class prior of an unlabeled set and derive 1C prior MMMF (1C-PMMMF). This method is motivated by the fact that in matrix factorization tasks (i.e. classification over two categorical domains) always a finite set of unlabeled (i.e. nonobserved) data points is present that contains all the possible examples of any test set. We assume there is some knowledge about the general class distribution and include this as a regularizer. This prior value can be either given by a domain expert or handled as a hyperparameter. We compare our class prior regularization approach to the simpler baselines of class prior thresholding.

In total our contributions are:

1. We transfer the idea of 1C-SVM to 1C-MMMF and show that here the size of the bias is invariant both under optimization and prediction.
2. 1C-MMMF is extended by class prior regularization to 1C-PMMMF.
3. An optimization algorithm for 1C-PMMMF that scales to large problems is developed.
4. We empirically show that 1C-PMMMF can outperform 1C-SVM in quality and can scale to problems that are infeasible for 1C-SVM.

10.2 Related Work

Recommender Systems

Item prediction from implicit feedback is similar to the problem we address here. The parallels are that also a binary relation over two high dimensional categorical domains is predicted and only positive feedback is given. The difference between matrix factorization techniques for the recommendation task (Hu et al, 2008; Pan et al, 2008) and our work is that they impute all non-observed values with negative class labels. That means that their approach assumes that the prior class value on the test set is 0. In contrast to this, we regularize the average binary predictions of an unlabeled set to a given class prior.

Outlier Detection & Semi-Supervised Learning

Binary classification using only feedback of one class is related to outlier detection (Hawkins, 1980) where the instances of the second (unobserved) class are the outliers. In 1C-PMMMF also information about the class distribution on a second set –most importantly the test set– is taken into account which makes it a semi-supervised method since knowledge about the unlabeled test data is used during learning. Similar to (Belkin et al, 2006) we exploit distributional properties of the unlabeled data (in our case: the prior class distribution) for regularization. In contrast to them we do not use geometric information since our task does not provide such information. In all, 1C-PMMMF can be seen as a semi-supervised outlier detection method (Hodge and Austin, 2004) for the extreme case where only one class is observed and the task is to define boundaries between this standard class and the outlier class using information about the prior class distribution on the unobserved data points.

One-Class Support Vector Machines

1C-SVMs (Schölkopf et al, 2001) are classifiers over numerical feature vectors. The standard approach to apply them to categorical variables is to use one indicator variable per level. For the Netflix example in our evaluation, that means there are 412,814 variables (for each of the 395,063 users and 17,751 items one variable) and every of the 36,915,512 cases is sparse as for each case only 2 variables are 1 and the remaining 412,812 variables are 0. In our evaluation, we use libSVM (Chang and Lin, 2001) that supports a sparse format.

10.3 One-Class Problems

One class problems are (binary) classification problems where only instances of one class are observed. Usually, the reason for observing only one class is the selection process of training examples, e.g. the system is only able to collect positive feedback. For example in Facebook there is only a 'like'-button but no 'hate'-button. In this chapter, we investigate one-class problems over two categorical variables ($m = 2$) with many levels. For easier readability, we refer to the first domain (X_1) as I and to the second one (X_2) as J. The task is to predict if a pair $(i, j) \in I \times J$ is positive or negative – e.g. if a user i loves a song j. All our problems are extremely sparse which means that only a small fraction of $I \times J$ is observed and of interest. Note that sparseness in this context means missing values and not that values are zeros.

10.3.1 Terminology

The following terminology is used: $I := \{i_1, \dots, i_m\}$ and $J := \{j_1, \dots, j_n\}$ are sets of objects. $Y : I \times J \to \{-1, 1\}$ is the target relation or equivalently Y is a binary $m \times n$ matrix: $Y : \{-1, 1\}^{m \times n}$. Depending on the context, we switch between both notations: $y_{i,j} = y(i, j)$. $S \subseteq I \times J$ is the set of observed pairs. In the one-class problems, we assume without loss of generality that only positive pairs are observed.

10.3.2 Prior Information

Even though, in one-class problems no training instances of negative classes are available, it is assumed that there might be negative examples in the future test set – otherwise running a classifier would not make much sense. Here, we investigate scenarios where a prior probability $p_D \in [0,1]$ on the class distribution on some set $D \subseteq I \times J$ is available. Some special cases of D are:

- $D = I \times J$: p_D is given on all data. As I and J are finite sets, also the set of all pairs is finite. Note that this differs from standard classification scenarios from real-valued feature data, where the set of all possible realizations is infinite.
- $D = (I \times J) \setminus S$: p_D is given on all unlabeled data. As the prior on S for one-class problems is $p_S = 1$, prior information $p_{D'}$ on $D' = I \times J$ can be transformed to prior information on $D = (I \times J) \setminus S$ and vice versa.
- $D = S_{\text{test}}$: p_D is given on all test examples. Knowing the test examples (but not their individual class label) makes it similar to semi-supervised learning.

In total, all three are scenarios where unlabeled data is taken into account.

10.4 One-Class Matrix Factorization

We start with introducing 2C-MMMF (Srebro et al, 2005) and 1C-SVM (Schölkopf et al, 2001). Then, we combine both ideas to 1C-MMMF and show the invariance to the bias. To solve this issue, we extend 1C-MMMF by a regularizer over the class prior of unlabeled data (1C-PMMMF). Then we compare this to a simpler approach of class prior thresholding.

Matrix Factorization (MF)

The matrix factorization model relies on approximating a partially observed matrix Z by the matrix product of two lower dimensional matrices:

$$\hat{Z} := W \cdot H^T \tag{10.1}$$

or equivalently:

$$\hat{z}_{i,j} := \langle w_i, h_j \rangle \tag{10.2}$$

with factorization matrices:

$$W \in \mathbb{R}^{m \times k}, \quad H \in \mathbb{R}^{n \times k}, \tag{10.3}$$

where k is the number of factorization dimensions.

As described in chapter 5, matrix factorization is the two-mode equivalent of PARAFAC and PITF. In contrast to the notation in chapter 5, we use $W := V^I$ and $H := V^J$ as feature matrices. The reason is, that we deal with a two-mode problem and here getting rid of the indices enhances readability.

Table 10.1 Prediction and optimization for 2C-MMMF (Srebro et al, 2005) and 1C-SVM (Schölkopf et al, 2001).

Method	Prediction	Optimization
2C-SVM	$\hat{y}(x) = \operatorname{sgn}(\langle w, \phi(x) \rangle)$	$\underset{w}{\operatorname{argmin}} \frac{1}{2}\|\|w\|\|^2 + \frac{1}{\nu l}\sum_{i=1}^{l} \max(0, 1 - y_i \langle w, \phi(x_i) \rangle)$
1C-SVM	$\hat{y}(x) = \operatorname{sgn}(\langle w, \phi(x) \rangle - \rho)$	$\underset{w,\rho}{\operatorname{argmin}} \frac{1}{2}\|\|w\|\|^2 + \frac{1}{\nu l}\sum_{i=1}^{l} \max(0, \rho - \langle w, \phi(x_i) \rangle) - \rho$
2C-MMMF	$\hat{y}_{i,j} = \operatorname{sgn}(\langle w_i, h_j \rangle)$	$\underset{W,H}{\operatorname{argmin}} \frac{1}{2}(\|\|W\|\|_F^2 + \|\|H\|\|_F^2) + \frac{1}{\nu \cdot \|S\|} \sum_{(i,j)\in S} \max(0, 1 - y_{i,j}\langle w_i, h_j \rangle)$

2C-MMMF

The classification rule of Maximum Margin Matrix Factorization for binary classification (2C-MMMF) (Srebro et al, 2005) and its optimization criterion can be found in table 10.1.

1-Class Support Vector Machines (SVM)

As SVM are classifiers over real valued vectors, we use the following notation: there are observations $\{(x_1, y_1), \ldots (x_l, y_l)\}$ where for each observation $(x, y) \in (\mathbb{R}^n \times \{-1, 1\})$. Let $\phi : \mathbb{R}^n \to \mathscr{F}$ be a mapping from the original feature space to the space $\mathscr{F}$.

Table 10.1 shows the optimization criterion and prediction formula of 1C-SVM and 2C-SVM. In contrast to normal 2C-SVM, in 1C-SVM the prediction function is shifted by a bias ρ: And the optimization criterion differs from 2C-SVM by the additional maximization w.r.t. ρ.

10.4.1 One-Class Maximum Margin Matrix Factorization (1C-MMMF)

To apply the ideas of 1C-SVM to MMMF, the classification rule of 1C-MMMF is biased with $\rho \in \mathbb{R}^+$ to the negative class:

$$\hat{y}_{i,j} = \operatorname{sgn}(\langle w_i, h_j \rangle - \rho) \tag{10.4}$$

Now, the optimization task is to maximize the bias like in 1C-SVM and simultaneously minimize the classification loss and the norm of W and H like in 2C-MMMF:

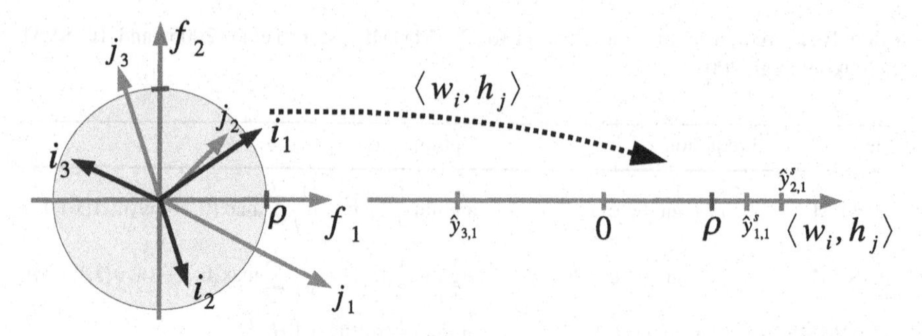

Fig. 10.1 1C-MMMF finds for each categorical variable a representation in $\mathbb{R}^k$ – here k=2 (see left side). The correlation $\langle w_i, h_j \rangle$ of the representation of a pair (i, j) has to exceed a bias ρ to be classified as positive (see right side). 1C-MMMF tries to minimize the overall length of all representations and to maximize the number of positive classifications ($\hat{y}_{i,j} = 1$) for the observed positive elements $(i, j) \in S$ – here $S = \{(1,1),(2,1)\}$.

$$\operatorname*{argmin}_{W,H,\xi,\rho} \frac{1}{2}(||W||_F^2 + ||H||_F^2) + \frac{1}{\nu \cdot |S|} \sum_{(i,j) \in S} \xi_{i,j} - \rho$$

subject to

$$\begin{aligned} &\forall (i,j) \in S : \langle w_i, h_j \rangle \geq \rho - \xi_{i,j}, \quad \xi_{i,j} \geq 0 \\ \Leftrightarrow &\forall (i,j) \in S : \xi_{i,j} \geq \rho - \langle w_i, h_j \rangle, \quad \xi_{i,j} \geq 0 \\ \Leftrightarrow &\forall (i,j) \in S : \xi_{i,j} \geq \max(0, \rho - \langle w_i, h_j \rangle) \end{aligned}$$

where $\xi_{i,j}$ are slack variables. Obviously this function is minimal if we have equality:

$$\operatorname*{argmin}_{W,H,\rho} \frac{1}{2}(||W||_F^2 + ||H||_F^2) + \frac{1}{\nu \cdot |S|} \sum_{(i,j) \in S} \max(0, \rho - \langle w_i, h_j \rangle) - \rho \tag{10.5}$$

This means, a 'simple' (short) representation W for elements in I and H for elements in J should be found such that the dot product of observed pairs $(i, j) \in S$ exceeds the classification threshold ρ (see figure 10.1). Until now, we have applied the same steps as in 1C-SVM to derive 1C-MMMF. Next, we show that for 1C-MMMF the particular choice of ρ is unimportant as long as ρ is positive and not equal to zero[1].

10.4.1.1 Scaling Invariance of the Bias ρ

The reason of the invariance of 1C-MMMF to the choice of the bias is that both the loss (eq. (10.5)) and the prediction formula (eq. (10.4)) can be scaled whereas in 1C-SVM this is not possible.

[1] It is important to note, that this does not mean, that the bias ρ can be skipped completely but only that the size/ length is unimportant – e.g. $\rho = 1$ can be chosen.

Lemma 10.1 (Loss under Scaling). *For any positive constant c, the loss is proportional to its rescaled variant:*

$$\forall c \in \mathbb{R}^+ : \quad cL(W,H,\rho) = L(\sqrt{c}W, \sqrt{c}H, c\rho)$$

Proof

$$\begin{aligned}
&L(\sqrt{c}W, \sqrt{c}H, c\rho) \\
=&\frac{1}{2}(||\sqrt{c}W||_F^2 + ||\sqrt{c}H||_F^2) - c\rho + \frac{1}{\nu \cdot |S|} \sum_{(i,j)\in S} \max(0, c\rho - \langle \sqrt{c}w_i, \sqrt{c}h_j \rangle) \\
=&\frac{c}{2}(||W||_F^2 + ||H||_F^2) - c\rho + \frac{c}{\nu \cdot |S|} \sum_{(i,j)\in S} \max(0, \rho - \langle w_i, h_j \rangle) \\
=&cL(W,H,\rho)
\end{aligned}$$

Lemma 10.2 (Prediction under Scaling). *The parameter settings* (W,H,ρ) *and* $(\sqrt{c}W, \sqrt{c}H, c\rho)$ *are equivalent under prediction.*

Proof

$$\begin{aligned}
\forall (i,j) \in I \times J : \hat{y}_{i,j}(\sqrt{c}W, \sqrt{c}H, c\rho) &= \operatorname{sgn}(\langle \sqrt{c}w_i, \sqrt{c}h_j \rangle - c\rho) \\
&= \operatorname{sgn}(c\langle w_i, h_j \rangle - c\rho) = \operatorname{sgn}(\langle w_i, h_j \rangle - \rho) \\
&= \hat{y}_{i,j}(W,H,\rho)
\end{aligned}$$

From both lemmas follows, that for any two biases $\rho_1 \in \mathbb{R}^+$ and $\rho_2 = c\rho_1 \in \mathbb{R}^+$, there are parameters (W_1^*, H_1^*) for ρ_1 and (W_2^*, H_2^*) for ρ_2 minimizing the loss in eq. (10.5) that are equivalent under prediction. Due to this scaling invariance of the solutions, ρ can be set to an arbitrary positive value. In the following, we fix $\rho = 1$. This simplifies the optimization criterion to:

$$\operatorname*{argmin}_{W,H} \frac{1}{2}(||W||_F^2 + ||H||_F^2) + \frac{1}{\nu \cdot |S|} \sum_{(i,j)\in S} \max(0, 1 - \langle w_i, h_j \rangle) \tag{10.6}$$

Comparing this to the optimization of 2C-MMMF (table 10.1) one can see that both are equal when only positive training instances are used for 2C-MMMF[2]. That means one can optimize 1C-MMMF problems with a 2C-MMMF solver to find W and H.

But optimizing (eq. (10.6)) with only one-class data will lead to a degenerate solution requiring only one latent dimension (see figure 10.3). The reason is that both feature matrices W and H are free in contrast to the 1C-SVM case where the feature matrix X is given. That means there is no need for (eq. (10.6)) to distinguish between feature dimensions because only positive training data is given. The Tikhonov regularization alone does not solve this problem because it just tries to

[2] Note that even though the optimization criterion is equivalent, the prediction of 1C-MMMF (eq. (10.4)) is different from 2C-MMMF (table 10.1) as it has a bias $\rho = 1$.

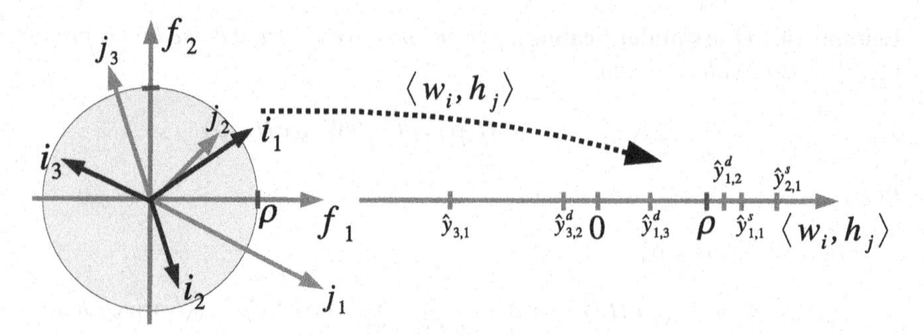

Fig. 10.2 1C-PMMMF takes prior knowledge about the class distribution p_D on unlabeled data D (here $\{(1,2),(1,3),(3,2)\}$) into account. 1C-PMMMF tries additionally to find representations such that the average classification of all values in D is close to p_D. E.g. for $p_D = 0.66$, 1C-PMMMF might try to find different feature vectors (left side) so that also $\hat{y}_{1,3} > \rho$ (right side).

minimize W and H. That is why, we will introduce a more selective regularization that demands negative classifications on unlabeled data which leads to an identification of factorization dimensions.

10.4.2 Class Prior Regularization

1C-MMMF is extended by a regularization term that makes use of additional information on the class prior distribution p_D on an unlabeled set D (see section 10.3.2). Thus, the model is called one-class **prior** MMMF (1C-PMMMF).

First, we define the estimated class prior $\hat{p}_D$ of the classifier $\hat{y}$ on $D \subset I \times J$ as:

$$\hat{p}_D := \frac{1}{|D|} \sum_{(i,j)\in D} \left(\frac{1}{2}\hat{y}_{i,j} + \frac{1}{2} \right) \tag{10.7}$$

Note that $\hat{p}_D$ is an average over binary class values ($\hat{y}_{i,j} \in \{-1,1\}$).

Now, we add an additional regularizer to the optimization that punishes deviations of the predicted class prior $\hat{p}_D$ from the prior knowledge p_D (see figure 10.2). This results in the following optimization task[3]:

$$\operatorname*{argmin}_{W,H,\xi,\zeta} \frac{1}{2}(||W||_F^2 + ||H||_F^2) + \frac{1}{\nu \cdot |S|} \sum_{(i,j)\in S} \xi_{i,j} - \rho + \frac{\rho}{\mu} \zeta$$

subject to

[3] We have chosen to make the additional prior constraint ζ linearly dependent on ρ like all other parameters are (see section 10.4.1.1).

$$\forall (i,j) \in S: \langle w_i, h_j \rangle \geq \rho - \xi_{i,j}, \quad \xi_{i,j} \geq 0$$
$$\zeta \geq (\hat{p}_D - p_D)^2$$

where $\mu \in \mathbb{R}^+$ controls the importance of the error of the prior estimate. Again, $\xi_{i,j}$ are the slack variables.

It is easy to show that all the arguments for the invariance of the loss to ρ for 1C-MMMF are also met here and so, the final optimization task is:

$$\underset{W,H}{\operatorname{argmin}} \frac{1}{2}(||W||_F^2 + ||H||_F^2) + \frac{1}{\nu \cdot |S|} \sum_{(i,j)\in S} \max(0, 1 - \langle w_i, h_j \rangle) + \frac{1}{\mu}(\hat{p}_D - p_D)^2$$

Please note that the class prior regularizer does **not** try to give each $\hat{y}_{i,j}$ a specific value (e.g. p_D) instead it makes a less restrictive statement about the average classification of all pairs $(i,j) \in D$.

In the following, we will use an equivalent formulation of this optimization formula that makes the trade-off between the three terms (max-margin regularization, classification loss and class prior regularization) more obvious. Instead of ν and μ, we will use the two hyperparameters η and λ:

$$\eta := \frac{\mu}{\mu + \nu}, \quad \lambda := \frac{\mu \nu}{\mu + \nu}$$

η can be interpreted as how much the classification loss should be favored over the class prior regularization. $\eta \in [0,1]$ because $\mu \in \mathbb{R}^+$ and $\nu \in \mathbb{R}^+$. The hyperparameter λ can be interpreted as how strong the maximum margin regularization term should be. With these definitions, the minimization task is:

$$\underset{W,H}{\operatorname{argmin}} \lambda \frac{1}{2}(||W||_F^2 + ||H||_F^2) + \frac{\eta}{|S|} \sum_{(i,j)\in S} \max(0, 1 - \langle w_i, h_j \rangle) + (1-\eta)(\hat{p}_D - p_D)^2 \tag{10.8}$$

10.4.3 Class Prior Thresholding

A simpler approach than using p_D for regularization is to use it as a threshold for post-processing. Given any scoring function $\hat{z}: I \times J \rightarrow \mathbb{R}$ and a prior probability p_D on a set D, we can create a simple threshold method that classifies all values exceeding the threshold θ_D as positive and the rest as negative:

$$\hat{y}_{i,j} := \operatorname{sgn}(\hat{z}_{i,j} - \theta_D) \tag{10.9}$$

The value of θ_D can be found by ordering the elements of D by $\hat{z}$ and then the value of θ_D is the one where $p_D \cdot |D|$ values are larger. This is the value minimizing $(\hat{p}_D - p_D)^2$.

Every estimator $\hat{z}$ can be used with this threshold method. Among these are:

1C-MMMF

The difference between 1C-MMMF with prior thresholding to prior regularization (1C-PMMMF) is that with thresholding the prior information is only used after the factors are learned. That means only the one-dimensional decision boundary (threshold) can be shifted, whereas 1C-PMMMF can use the prior information to learn other factorizations. Our evaluation shows that this is an important difference. Furthermore, 1C-MMMF with prior thresholding and 2C-MMMF with prior thresholding are equivalent because (i) we already have shown that the optimization is equivalent and (ii) now also the prediction gets the same as both are shifted ($\theta_D^{\text{2C-MMMF}} = \theta_D^{\text{1C-MMMF}} + 1$).

Row/ Column Estimators

An even simpler estimator for scoring a pair (i, j) is to count how often other elements in the same row i are positive. This might be done either row-wise or column-wise or both can be linearly ensembled:

$$\hat{z}_{i,j} := |\{(i, j') \in S\}| + |\{(i', j) \in S\}|$$

k-Nearest Neighbour

This method weights the entries either over rows or columns using a similarity measure (e.g. cosine similarity):

$$\hat{z}_{i,j} := \sum_{(i,j') \in S} \frac{|\{i' | (i', j) \in S \wedge (i', j') \in S\}|}{\sqrt{|\{i' | (i', j) \in S\}| \cdot |\{i' | (i', j') \in S\}|}}$$

10.4.4 Scalable Learning

In the following, we will present a scalable method for learning a 1C-PMMMF model. As 1C-MMMF is a special case of 1C-PMMMF ($\eta = 1$), this algorithm also covers the simpler model class. Our proposed algorithm is based on stochastic gradient descent. Applying typical stochastic gradient descent to eq. (10.8) is difficult, as there are two pools of cases: S and D. Our approach is to split the minimization criterion into two losses and alternate between them (see algorithm 7).

First we split the loss of 1C-PMMMF (eq. (10.8)) into:

$$L(W, H) = \eta L_C(W, H) + (1 - \eta) L_P(W, H)$$

Algorithm 7 LEARN1CPMMMF

Input: training data S, prior information p_D on D, learning rate α
Output: model parameters W, H

1: initialize W, H
2: **repeat**
3: draw $x_\eta \sim U(0,1)$
4: **if** $x_\eta \leq \eta$ **then**
5: draw (i, j) uniformly from S
6: **for** $f \in \{1, \ldots, k\}$ **do**
7: $w_{i,f} \leftarrow w_{i,f} - \alpha \frac{\partial}{\partial w_{i,f}} L_C(W,H)$
8: $h_{j,f} \leftarrow h_{j,f} - \alpha \frac{\partial}{\partial h_{j,f}} L_C(W,H)$
9: **end for**
10: **else**
11: draw (i, j) uniformly from D
12: **for** $f \in \{1, \ldots, k\}$ **do**
13: $w_{i,f} \leftarrow w_{i,f} - \alpha \frac{\partial}{\partial w_{i,f}} L_P(W,H)$
14: $h_{j,f} \leftarrow h_{j,f} - \alpha \frac{\partial}{\partial h_{j,f}} L_P(W,H)$
15: **end for**
16: **end if**
17: **until** stopping criterion is met
18: **return** W, H

where

$$L_C(W,H) := \lambda \frac{1}{2}(||W||_F^2 + ||H||_F^2) + \frac{1}{|S|} \sum_{(i,j) \in S} \max(0, 1 - \langle w_i, h_j \rangle)$$

$$L_P(W,H) := \lambda \frac{1}{2}(||W||_F^2 + ||H||_F^2) + (\hat{p}_D - p_D)^2$$

Because $\eta \in [0,1]$, we propose to alternate between optimizing L_C and L_P by drawing cases (i, j) from S and D respectively where the probability of drawing from S is η.

Given a case $(i, j) \in S$, the gradients for L_C are:

$$\frac{\partial}{\partial w_{i,f}} L_C(W,H) = \lambda\, w_{i,f} + \begin{cases} 0, & \text{if } \langle w_i, h_j \rangle > 1 \\ -h_{j,f}, & \text{else} \end{cases}$$

$$\frac{\partial}{\partial h_{j,f}} L_C(W,H) = \lambda\, h_{i,f} + \begin{cases} 0, & \text{if } \langle w_i, h_j \rangle > 1 \\ -w_{i,f}, & \text{else} \end{cases}$$

where f is the index over the factorization features. If the case is drawn from D, the gradients for L_P are:

$$\frac{\partial}{\partial w_{i,f}} L_P(W,H) = \lambda\, w_{i,f} + (\hat{p}_D - p_D) \frac{\partial}{\partial w_{i,f}} \hat{y}_{i,j}$$
$$\frac{\partial}{\partial h_{j,f}} L_P(W,H) = \lambda\, h_{j,f} + (\hat{p}_D - p_D) \frac{\partial}{\partial h_{j,f}} \hat{y}_{i,j}$$

As $\hat{y}_{i,j} = \text{sgn}(\langle w_i, h_j \rangle - 1)$ is not differentiable, we replace $\text{sgn}(x)$ with the logistic function $\sigma(x) = \frac{1}{1+e^{-x}}$. To ensure same scales as $\text{sgn}(x)$, we use the scaled sigmoid function $2\sigma(x) - 1$. With this definition, the smooth gradients of $\hat{y}_{i,j}$ are approximated by:

$$\frac{\partial}{\partial w_{i,f}} \hat{y}_{i,j} \approx 2 h_{j,f}\, \sigma(\langle w_i, h_j \rangle - 1)\,(1 - \sigma(\langle w_i, h_j \rangle - 1))$$
$$\frac{\partial}{\partial h_{j,f}} \hat{y}_{i,j} \approx 2 w_{i,f}\, \sigma(\langle w_i, h_j \rangle - 1)\,(1 - \sigma(\langle w_i, h_j \rangle - 1))$$

The learning rate α can easily be found by trying several values and testing the convergence of L for each choice. The computational complexity of the described algorithm is $O(qkC(\hat{p}_D)\,|S|)$ – where q is the number of iterations[4], k the factorization dimension and $C(\hat{p}_D)$ the complexity of computing $\hat{p}_D$. As the problems are usually very sparse, S is typically much smaller than $m \cdot n$ – see the example in our evaluation. The only bottleneck so far is the calculation of $\hat{p}_D$ because it is defined as the expectation value over the classification over all elements in D and thus its exact computation complexity is in $O(|D|)$. Thus in the following, we will describe a way of estimating $\hat{p}_D$ in $O(1)$ which leads to an overall complexity of $O(qk\,|S|)$ for our proposed learning algorithm – i.e. it is linear in the number of observations $|S|$ and the dimensionality of the factorization k.

10.4.4.1 Calculation of $\hat{p}_D$

As the changes in $\hat{p}_D$ after a single update step are small, one can recalculate $\hat{p}_D$ only after a certain number of iterations (e.g. 1000). Secondly, if D is large, a subsample $D^* \subset D$ can be used to estimate $\hat{p}_D$. We suggest to draw D^* uniformly from D and use $\hat{p}_{D^*}$ to estimate $\hat{p}_D$.

Lemma 10.3 (Statistical properties of $\hat{p}_{D^*}$). *Based on uniformly (with replacement) sampled instances (i, j) from D the estimator $\hat{p}_{D^*} := \frac{1}{|D^*|} \sum_{D^*} \frac{\hat{y}_{i,j}+1}{2}$ is an unbiased estimator for $\hat{p}_D$ with variance $Var(\hat{p}_{D^*}) = \frac{\hat{p}_D(1-\hat{p}_D)}{|D^*|}$*

Since D^* is a uniformly sampled subset of the population D of stochastically independent binary random variables $\hat{y}_{i,j}$ for different pairs (i, j)[5] unbiasedness and variance result from elementary probability theory. Note that the probability $p(\hat{y}_{i,j} = 1)$

[4] We define an iteration as performing $|S|$ stochastic updates.

[5] Note that $\hat{y}_{i,j}$ is a random variable defined on pairs (i, j). Therefore this random variable is independent for any other pair (k, l) where at least one of the two indices is different (i.e. $i \neq k$ or $j \neq l$).

Table 10.2 Characteristics of the evaluated datasets.

Dataset	n (user)	m (items)	$\lvert S\rvert$ (events)
Netflix	395,063	17,751	36,915,512
Movielens	6,040	3,952	383,972

of sampling a positive variable is $\hat{p}_D$. Moreover, $\hat{p}_{D^*}$ is asymptotically normal distributed with mean $\mu = \hat{p}_D$ and variance $\sigma^2 = \frac{1}{|D^*|}\hat{p}_D(1-\hat{p}_D)$. Hence, subsampling the data (e.g. $|D^*| = 1000$) reduces the time-complexity of the proposed learning algorithm considerably to constant time while the thereof provoked additional error is neglectable and independent of $|D|$.

10.5 Evaluation

Next, we compare 1C-PMMMF to 1C-SVM, 1C-MMMF and the prior-based threshold methods k-nearest-neighbor ('θ_p kNN') and an ensemble over row- and column-weighting ('θ_p Ensemble'). We also apply the prior-threshold method to 1C-MMMF ('θ_p 1C-MMMF') which is the same as ('θ_p 2C-MMMF') – see the discussion in section 10.4.3.

10.5.1 Dataset and Methodology

We convert the Netflix[6] and Movielens[7] datasets into a binary prediction problem of 'hating' (1 or 2 stars) or 'loving' (5 stars) a movie. Afterwards, we create a 10-core subset, i.e. users with less than 10 ratings and movies with less than 10 ratings are removed. The characteristics can be found in table 10.2.

The datasets are split into a training set S and test set $S_{\text{test}} =: D$ of same size and all observations with the negative class label are removed from the training set. We repeat all experiments 5 times with new training/ test splits and report the mean accuracy. Note that our experiments are quite extensive as more than 320 different factorization models had to be trained; half of them on the large-scale Netflix dataset. For 1C-SVM, 1C-PMMMF and 1C-MMMF, we search for the best regularization ν and λ respectively on the first fold and use this on the remaining folds.

10.5.2 Results

Figure 10.3 compares all methods with varying hyperparameters. With a reasonable factorization dimension and prior estimate, 1C-PMMMF outperforms the other methods.

[6] `http://www.netflixprize.com/`
[7] `http://www.grouplens.org/`

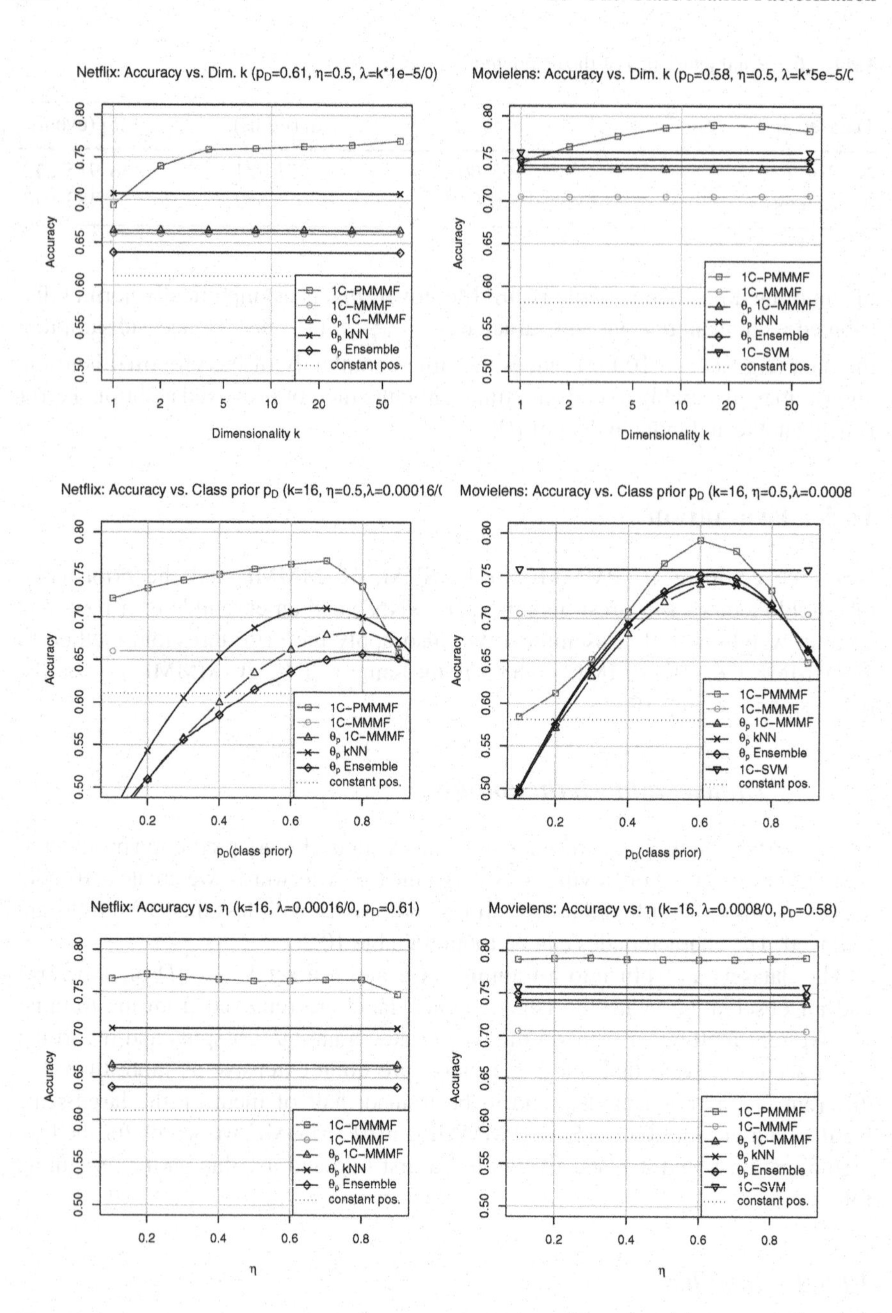

Fig. 10.3 Comparison of 1C-PMMMF with class prior regularization to 1C-MMMF, 1C-SVM (Schölkopf et al, 2001) and class prior thresholding for the methods 1/2C-MMMF, kNN and a row/ column ensemble. With a reasonable chosen prior, 1C-PMMMF outperforms all other methods in quality. Note that we cannot state results for 1C-SVM on Netflix because training did not converge even after 100 days of runtime.

10.5.2.1 Class Prior Regularization vs. Thresholding

Comparing class prior regularization (1C-PMMMF) to thresholding (θ_p methods), one can see that 1C-PMMMF outperforms all thresholding methods for almost any choice of hyperparameters. Especially this shows that for matrix factorization, class prior regularization (1C-PMMMF) leads to a different (better) factorization than class prior thresholding. Furthermore, the results of 1C-MMMF for the dimensionality experiment matches our hypothesis (see section 10.4.1) that 1C-MMMF cannot benefit from more factorization dimensions because it cannot identify any 'semantics' in the dimensions. The class prior regularization solves this issue.

10.5.2.2 1C-PMMMF vs. 1C-SVM

Applying 1C-SVM with binary indicator variables leads to good results on the small Movielens dataset. But if the class prior is estimated good enough (here ca. ± 0.1), 1C-PMMMF gives better quality. For the larger Netflix dataset, 1C-SVM is not feasible and very slow. In contrast to this, training 1C-PMMMF with $k = 16$ takes only several hours on Netflix.

10.5.2.3 Sensitivity to Hyperparameters

All class prior methods are sensitive to how good the provided hyperparameter p_D matches to the real class prior (0.61 for Netflix, 0.58 for Movielens). On Netflix it is interesting to see that 1C-PMMMF seems to be much less sensitive than the thresholding methods. Another interesting result is that 1C-PMMMF is not sensitive to the choice of η within $[0.1, 0.9]$ – for $\eta = 1$ 1C-PMMMF corresponds to 1C-MMMF. Remember that η is used as alternating probability between class prior regularization and classification error minimization. Finally, both 1C-PMMMF and 1C-SVM are sensitive to the choice of λ and ν respectively. But with 1C-PMMMF good results are also obtained with $\lambda = 0$ provided that k is chosen small (e.g. 8-32).

10.6 Conclusion

We have introduced a matrix factorization method for one-class classification over two high-dimensional categorical variables. First, we have applied the ideas of 1C-SVM to obtain 1C-MMMF. Then we have argued that 1C-MMMF alone is not applicable as it is not able to identify the 'semantics' of any dimension. We therefore introduced a regularization term that is based on the class prior and show the difference to prior thresholding. Furthermore a scalable learning algorithm for 1C-PMMMF has been introduced that is able to solve large scale problems over thousands of variable levels and millions of observations where 1C-SVM is not applicable any more. The empirical results indicate that 1C-PMMMF is able to generate high quality predictions for one-class problems outperforming other methods like prior-threshold based methods or 1C-SVM.

In contrast to context-aware ranking from implicit feedback, we assume that the class distribution is not heavily biased to the non-observed (negative) class. That means, the one-class methods of this chapter are applicable for more or less balanced classification problems whereas the context-aware ranking methods introduced before (like item recommmendation, chapter 6) are suitable for heavily unbalanced problems, where the target is ranking instead of classification.

References

Belkin, M., Niyogi, P., Sindhwani, V.: Manifold regularization: A geometric framework for learning from labeled and unlabeled examples. Journal of Machine Learning Research 7, 2399–2434 (2006)

Chang, C.C., Lin, C.J.: LIBSVM: a library for support vector machines (2001), Software available at http://www.csie.ntu.edu.tw/~cjlin/libsvm

Hawkins, D.: Identification of Outliers. Chapman and Hall, Boca Raton (1980)

Hodge, V., Austin, J.: A survey of outlier detection methodologies. Artif. Intell. Rev. 22(2), 85–126 (2004)

Hu, Y., Koren, Y., Volinsky, C.: Collaborative filtering for implicit feedback datasets. In: IEEE International Conference on Data Mining (ICDM 2008), pp. 263–272 (2008)

Pan, R., Zhou, Y., Cao, B., Liu, N.N., Lukose, R.M., Scholz, M., Yang, Q.: One-class collaborative filtering. In: IEEE International Conference on Data Mining (ICDM 2008), pp. 502–511 (2008)

Rennie, J.D.M., Srebro, N.: Fast maximum margin matrix factorization for collaborative prediction. In: ICML 2005: Proceedings of the 22nd International Conference on Machine Learning, pp. 713–719. ACM, New York (2005)

Schölkopf, B., Platt, J.C., Shawe-Taylor, J., Smola, A.J., Williamson, R.C.: Estimating the support of a high-dimensional distribution. Neural Computation 13(7), 1443–1471 (2001)

Srebro, N., Rennie, J.D.M., Jaakola, T.S.: Maximum-margin matrix factorization. In: Advances in Neural Information Processing Systems, vol. 17, pp. 1329–1336. MIT Press, Cambridge (2005)

Part V
Conclusion

Chapter 11
Conclusion

In this book, we have studied multi-mode prediction problems. The focus was on problem settings with large categorical domains and high sparsity. Due to the sparsity and usually high imbalance, we are not interested in classification but in ranking the entities of one of the modes. Instead of creating one global ranking, the rankings should be context-aware – i.e. we create many rankings that depend on a given context. Important applications for this setting are recommender systems. There exist several recommender tasks, two of the most well studied ones are personalization and tag recommendation. Our developed method of context-aware ranking subsumes both of them and includes also other settings like time-awareness. Moreover other well-known applications like web search or multi-label classification (e.g. annotation, wikipedia categorization) can be seen as an instance of context-aware ranking.

Based on these ideas, we have introduced the theory of context-aware ranking. From a Bayesian analysis, we have derived the optimization criterion BCR-OPT which is the general MAP estimator of a parametrized model. The learning algorithm BCR-LEARN is a maximization procedure for BCR-OPT. As it is based on stochastic gradient descent, any model that can be expressed as a differentiable, non-recursive function with a finite set of parameters can be optimized. The bootstrapping approach with the proposed drawing schemes makes it applicable even in cases where the number of ranking triples is huge. We have demonstrated the usefulness of BCR on the task of item recommendation, tag recommendation and sequential set recommendation. Here, we have applied BCR to a wide variety of models including several factorization models (matrix factorization, TD, PARAFAC, PITF), k-nearest-neighbour and Markov chains. Throughout these experiments, BCR optimization has outperformed other state-of-the-art approaches like weighted regularized least-square in quality. These results indicate that choosing the right optimization criterion is important. Furthermore, it shows that BCR is generic and a good choice for many applications.

With respect to modelling, we have focused on factorization models. For two-mode problems, matrix factorization models are known to generate high quality predictions (e.g. for regression or binary classification). In this book, we extended

S. Rendle: Context-Aware Ranking with Factorization Models, SCI 330, pp. 173–176.
springerlink.com

these factorization approaches to multi-mode problems. The Tucker decomposition and the PARAFAC model are such multi-mode factorization models. We have discussed their strength and limitations in detail showing that TD results in slow runtime for our multi-mode settings. Furthermore our empirical results indicate that optimizing model parameters for both TD and PARAFAC with standard Gaussian priors (aka ridge regression) can result in bad prediction quality. Our assumption is that in sparse settings, both of these models are too expressive and lack an a priori structure. To solve this, we propose the more restricted model PITF that explicitly models pairwise-interactions. We have shown that both TD and PARAFAC subsume this model. But in our evaluation, PITF outperforms both TD and PARAFAC. Thus our empirical results indicate that restricting the expressiveness and predefining a structure on TD models makes sense in sparse settings.

General factorization models like TD assume finite variable domains. For handling infinite domains like time, we present two extensions. The first one is a Markov chain, where sequential pattern can be found. Here, we extend the general Markov chain with personalization and secondly we model the transition cube with a factorization model (e.g. TD/PARAFAC/PITF). We have shown that this model subsumes both standard MCs and the standard non-time-aware factorization models. The second kind of time-variance we investigate, is variance within factors. Each factor is modelled time-dependent by decomposing it into basis functions and free parameters. We have shown, how these basis functions can be generated/ sampled from the observed data using a kernel approach.

11.1 Summary of Contributions

In total, the contribution of this book are:

1. **Theory of Context-Aware Ranking**
   We introduce context-aware ranking. This subsumes many important tasks like item recommendation and tag recommendation. Although each of these tasks has attracted a lot of research, they are usually studied isolated of each other. Our work brings them together which allows to transfer results from one domain to the other.
2. **BCR optimization and learning**
   We develop the Bayesian Context-aware Ranking (BCR) method that consists of a new optimization criterion BCR-OPT and a learning algorithm BCR-LEARN. BCR-OPT is the MAP estimator of the model parameters given context-aware ranking constraints that are derived from sparse observations. BCR-LEARN is a generic learning algorithm for context-aware ranking that is based on stochastic gradient descent with bootstrap sampling.
3. **Factorization Models (PITF)**
   For modelling the multi-mode data, we suggest to use factorization models based on Tucker decomposition. We discuss the practical limitations of general Tucker decomposition in terms of runtime and regularization. To solve this, we have

developed the pairwise interaction model PITF and shown that for ranking problems, the model complexity is linear.

4. **Factorizing Personalized Markov Chains**
   Moreover, we have extended Markov chain models by personalization and factorization. Personalization allows each user to have an individual Markov chain – i.e. own transition probabilities. By factorizing transition cubes, we solve the problem of sparsity in the data. That means, information propagates over the whole transition cube and the estimation gets more reliable than MLE with full parametrized models.
5. **Empirical Studies & Applications**
   We have applied our theory of context-aware ranking, the BCR optimization method and the factorization models to several applications: (1) online shopping, movie rental and online TV (Gantner et al, 2009) (item recommendation), (2) bookmark and music tagging (tag recommendation) and (3) sequential basket recommendation. In all of these applications, our proposed models have shown to outperform current state-of-the art methods. Furthermore our method has won the tag recommender challenge of the 2009 ECML/PKDD Discovery Challenge.
6. **Time-aware Factor Model**
   For handling time variance, we have extended the general factorization model with time-variant factors. This is done by decomposing each factor into a set of basis functions and free parameters. We provide a method for generating the set of basis functions from the observed data using a kernel assumption.
7. **One-Class Matrix Factorization**
   We extend the binary Maximum Margin Matrix Factorization classifier to handle one-class problems. This is done by transferring ideas from one-class support vector machines to matrix factorization. We show that the optimization of 1C-MMMF is invariant to the size of the bias and thus can be kept constant. We extend this model by prior regularization to 1C-PMMMF which allows to take assumptions about the class distribution prior into account.

11.2 Future Directions

Besides the directions that have been mentioned for specific tasks throughout this work, we see three major directions of future work:

- **Multirelational Prediction**
  The settings of this work can be seen as single relational. That means that there is one relation (e.g. a customer buys a product) and this relation should be predicted – i.e. the instances of one variable should be ranked given the others. But often additional information is available like information about products (price, category, ...) or information about customers (gender, age, ...). Especially in sparse settings, this additional information might be helpful to create better predictions of the target relation. There is already much work (Tso and Schmidt-Thieme, 2006; Agarwal and Chen, 2009; Tso-Sutter et al, 2008) on special cases (e.g. attribute-aware or tag-aware recommender systems) but only limited one for the

general case (Lin et al, 2009). Thus, one direction of future work is to develop a generic method for multi-relational settings and to integrate it in our work on context-aware ranking.

- **Regularization**
  As the empirical results for tag recommendation have shown, the restricted PITF is able to outperform PARAFAC and Tucker decomposition. But both PARAFAC and TD subsume PITF, so they should be able to generate at least as good predictions as PITF. We have argued that the reason might be the regularization together with the sparsity. Thus, a very interesting topic would be to investigate other regularization methods than 0-based Gaussian priors. Solving this problem would help to build better factorization models where the structure of the model is chosen data-dependent.
- **Applications**
  In this book we could only investigate some applications of context-aware ranking. Due to the success of our approach in our selected applications, we assume that it might also improve results on other tasks. Examples for further applications are: (1) personalized web-search, where web pages are suggested for a user given a query, (2) context-aware advertising, where ads are ranked based on a context like user, web page, last actions, (3) annotation like wikipedia categorization or (4) multi-label classification in general.

References

Agarwal, D., Chen, B.C.: Regression-based latent factor models. In: KDD 2009: Proceedings of the 15th ACM SIGKDD International Conference on Knowledge Discovery and Data Mining, pp. 19–28. ACM, New York (2009)

Gantner, Z., Freudenthaler, C., Rendle, S., Schmidt-Thieme, L.: Optimal ranking for video recommendation. In: Personalization in Media Delivery Platforms Workshop at the International ICST Conference on User Centric Media (PerMeD 2009) (2009)

Lin, Y.R., Sun, J., Castro, P., Konuru, R., Sundaram, H., Kelliher, A.: Metafac: community discovery via relational hypergraph factorization. In: KDD 2009: Proceedings of the 15th ACM SIGKDD International Conference on Knowledge Discovery and Data Mining, pp. 527–536. ACM, New York (2009)

Tso, K., Schmidt-Thieme, L.: Evaluation of attribute-aware recommender system algorithms on data with varying characteristics. In: Ng, W.-K., Kitsuregawa, M., Li, J., Chang, K. (eds.) PAKDD 2006. LNCS (LNAI), vol. 3918, pp. 831–840. Springer, Heidelberg (2006)

Tso-Sutter, K., Marinho, L., Schmidt-Thieme, L.: Tag-aware recommender systems by fusion of collaborative filtering algorithms. In: Proceedings of 23rd Annual ACM Symposium on Applied Computing (SAC 2008), Fortaleza, Brazil (to appear) (2008)

Glossary

Definition	Description
$\|\|X\|\|_F^2 := \sum_{i,j} x_{i,j}^2$	Frobenius norm
$\delta(b) := \begin{cases} 1, & \text{if } b \text{ is true} \\ 0, & \text{else} \end{cases}$	Delta function
$\sigma(x) := \frac{1}{1+e^{-x}}$	Sigmoid/ logistic function
$\langle \mathbf{x}, \mathbf{y} \rangle := \sum_{i=1} x_i \cdot y_i$	Dot product

Index